Air Pollution and Health

ONE WEEK LOAN

3 1 MAY 2000

ISSUES IN ENVIRONMENTAL SCIENCE AND TECHNOLOGY

EDITORS:

R. E. Hester, University of York, UK
R. M. Harrison, University of Birmingham, UK

EDITORIAL ADVISORY BOARD:

Sir Geoffrey Allen, Executive Advisor to Kobe Steel Ltd, **A. K. Barbour,** Specialist in Environmental Science and Regulation, UK, **N. A. Burdett,** Eastern Electricity, UK, **J. Cairns, Jr.,** Virginia Polytechnic Institute and State University, USA, **P. A. Chave,** Water Pollution Consultant, UK, **P. Crutzen,** Max-Planck-Institut für Chemie, Germany, **S. J. de Mora,** Université de Québec à Rimouski, Canada, **P. Doyle,** Zeneca Group PLC, UK, **Sir Hugh Fish,** Consultant, UK, **M. J. Gittins,** Consultant, UK, **J. E. Harries,** Imperial College, London, UK, **P. K. Hopke,** Clarkson University, USA, **Sir John Houghton,** Royal Commission on Environmental Pollution, UK, **N. J. King,** Environmental Consultant, UK, **J. Lester,** Imperial College of Science, Technology and Medicine, UK, **S. Matsui,** Kyoto University, Japan, **D. H. Slater,** Task Force on Risk Assessment, UK, **T. G. Spiro,** Princeton University, USA, **D. Taylor,** Zeneca Limited, UK, **T. L. Theis,** Clarkson University, USA, **Sir Frederick Warner,** SCOPE Office, UK.

TITLES IN THE SERIES:

FORTHCOMING:

How to obtain future titles on publication

A subscription is available for this series. This will bring delivery of each new volume immediately upon publication. For further information, please write to:

The Royal Society of Chemistry
Turpin Distribution Services Limited
Blackhorse Road
Letchworth
Herts SG6 1HN, UK

Telephone: +44 (0) 1462 672555   Fax: +44 (0) 1462 480947

ISSUES IN ENVIRONMENTAL SCIENCE
AND TECHNOLOGY

EDITORS: R. E. HESTER AND R. M. HARRISON

10
Air Pollution and Health

THE ROYAL
SOCIETY OF
CHEMISTRY
Information
Services

ISBN 0-85404-245-8
ISSN 1350-7583

A catalogue record for this book is available from the British Library

© The Royal Society of Chemistry 1998

Published by The Royal Society of Chemistry, Thomas Graham House,
Science Park, Milton Road, Cambridge CB4 0WF, UK
For further information see our web site at www.rsc.org

Typeset in Great Britain by Vision Typesetting, Manchester
Printed and bound by Redwood Books Ltd., Trowbridge, Wiltshire

Preface

This volume complements Issue Number 8 on *Air Quality Management*. Together, the two represent a comprehensive overview of the major forces driving air pollution control and the policies and measures through which that air pollution control is achieved. In this volume on air pollution and health, some of the leading international figures in this field give their personal view on the current state of knowledge in the key areas in which air pollutants impact on the health of the general population. This has been a very active field in recent years, with a burgeoning of research activity and a drive by both national governments and international organizations such as the European Union and World Health Organization to set new health-based air quality standards based on the latest research designed to protect the public. Much of that recent activity is encapsulated in the chapters of this volume.

In the first article, Jon Ayres reviews the latest information on health impacts of the more important gaseous air pollutants. These include sulfur dioxide, nitrogen dioxide, carbon monoxide and ozone. Professor Ayres has been at the centre of activity in the UK designed to research and interpret data on the health impacts of air pollution, having chaired a government-appointed committee set the task of quantifying the public health impacts of air pollution exposure in the UK. The second article addresses the very topical issue of airborne particulate matter, widely measured as the fraction known as PM_{10}. Ken Donaldson and William MacNee have been at the forefront of research to understand the mechanisms by which PM_{10} causes lung injury, and their paper reviews research from around the world on this complex issue. It demonstrates clearly that inhaled particles can have profound impacts on physiology and biochemistry and it is therefore unsurprising that epidemiological studies show major impacts of PM_{10} exposure.

Public concern over pollution tends to focus very much on chemical carcinogens. The third article by John Christian Larsen and Poul Bo Larsen provides an extensive and authoritative review of the health effects of important airborne chemical carcinogens. By means of unit risk factors, very approximate estimates may be made of the public health impact of exposure to chemical carcinogens. This article serves well to put the issue into perspective. In the further article by Roy Harrison on setting health-based air quality standards, much of the earlier information is drawn together in an overall introduction to the methodology used to set standards, and to the standards available in Europe and North America. The methodology and approach used in standard setting depends very much on the nature of the pollutant, and this is brought out clearly by case studies of individual pollutants. The following article by Morton Lippmann deals specifically with the standards set for particulate matter and ozone by the US EPA in 1997. The North American standard setting process is both protracted and extremely thorough, and Professor Lippmann, who was personally involved in this process as well as participating in World Health Organization activities in the field, does a masterful job in summarizing the very complex issues within this article. The major part of the article deals with particulate matter and well complements the more mechanistic article by Donaldson and MacNee. Professor Lippmann concentrates especially on the epidemiological evidence and how that can be used to evaluate the risk from particulate matter exposure.

Air quality standards normally relate to outdoor air and the fact is often ignored that the air inside the home can be significantly more polluted than that outdoors. Whilst the mechanisms of effect are likely to be the same for indoor pollutants as for those outdoors, the pollutant mix is different, and quite independent studies have been made of the effects of indoor pollution. These are comprehensively reviewed by Paul Harrison, who draws his reviews of individual pollutants together into very helpful conclusions relating to each of the major pollutants. He also reviews policy and research initiatives.

We are delighted to have attracted such a distinguished group of authors in this very topical field. We feel that overall they have produced a superb overview of the subject, which will be of lasting value.

Roy M. Harrison
Ronald E. Hester

Contents

Contents

Editors

Ronald E. Hester, BSc, DSc(London), PhD(Cornell), FRSC, CChem

Ronald E. Hester is Professor of Chemistry in the University of York. He was for short periods a research fellow in Cambridge and an assistant professor at Cornell before being appointed to a lectureship in chemistry in York in 1965. He has been a full professor in York since 1983. His more than 300 publications are mainly in the area of vibrational spectroscopy, latterly focusing on time-resolved studies of photoreaction intermediates and on biomolecular systems in solution. He is active in environmental chemistry and is a founder member and former chairman of the Environment Group of The Royal Society of Chemistry and editor of 'Industry and the Environment in Perspective' (RSC, 1983) and 'Understanding Our Environment' (RSC, 1986). As a member of the Council of the UK Science and Engineering Research Council and several of its sub-committees, panels and boards, he has been heavily involved in national science policy and administration. He was, from 1991–93, a member of the UK Department of the Environment Advisory Committee on Hazardous Substances and is currently a member of the Publications and Information Board of The Royal Society of Chemistry.

Roy M. Harrison, BSc, PhD, DSc (Birmingham), FRSC, CChem, FRMetS, FRSH

Roy M. Harrison is Queen Elizabeth II Birmingham Centenary Professor of Environmental Health in the University of Birmingham. He was previously Lecturer in Environmental Sciences at the University of Lancaster and Reader and Director of the Institute of Aerosol Science at the University of Essex. His more than 250 publications are mainly in the field of environmental chemistry, although his current work includes studies of human health impacts of atmospheric pollutants as well as research into the chemistry of pollution phenomena. He is a past Chairman of the Environment Group of The Royal Society of Chemistry for whom he has edited 'Pollution: Causes, Effects and Control' (RSC, 1983; Third Edition, 1996) and 'Understanding our Environment: An Introduction to Environmental Chemistry and Pollution' (RSC, Second Edition, 1992). He has a close interest in scientific and policy aspects of air pollution, having been Chairman of the Department of Environment Quality of Urban Air Review Group as well as currently being a member of the DETR Expert Panel on Air Quality Standards and Photochemical Oxidants Review Group, the Department of Health Committee on the Medical Effects of Air Pollutants and Chair of the DETR Atmospheric Particles Expert Group.

Contributors

J. G. Ayres, *Heartlands Research Institute, Birmingham Heartlands Hospital, Bordesley Green East, Birmingham B9 5SS, UK*

K. Donaldson, *Biological Sciences, Napier University, 10 Colinton Road, Edinburgh EH10 5DT, UK*

P. T. C. Harrison, *MRC Institute for Environment and Health, University of Leicester, 94 Regent Road, Leicester LE1 7DD, UK*

R. M. Harrison, *Institute of Public and Environmental Health, University of Birmingham, Edgbaston, Birmingham B15 2TT, UK*

J. C. Larsen, *Institute of Food Safety and Toxicology, Danish Veterinary and Food Administration, 19 Mørkhoj Bygade, DK-2860 Søborg, Denmark*

P. B. Larsen, *Institute of Food Safety and Toxicology, Danish Veterinary and Food Administration, 19 Mørkhoj Bygade, DK-2860 Søborg, Denmark*

M. Lippmann, *Nelson Institute of Environmental Medicine, New York University Medical Center, 57 Old Forge Road, Tuxedo, NY 10987, USA*

W. MacNee, *Unit of Respiratory Medicine, University of Edinburgh, Teviot Place, Edinburgh EH8 9AG, UK*

Health Effects of Gaseous Air Pollutants

JON G. AYRES

1 Introduction

Gaseous air pollutants constitute an important overall component of both outdoor and indoor air and are recognized to cause health effects, essentially in individuals with pre-existing disease. For the purposes of this chapter the gases, the primary pollutants sulfur dioxide (SO_2), nitrogen dioxide (NO_2) and carbon monoxide (CO) with the secondary pollutant ozone, will be considered whereas acidic species will not as they are generally regarded as part of the particulate fraction. As it is likely to be the acidic nature of those species that are important in health terms, even those acids which are present in the air as a vapour phase will not be considered here.

The sources of these pollutants are important when considering health effects because sources relate to individual and population exposures. The main source of SO_2 is from fossil fuel burning, the major contributors in the UK being coal-fired power stations. Nitrogen dioxide is derived from vehicle emissions, industrial sources (including power stations) and, in the indoor environment, from combustion of gas. Although for smokers of cigarettes the major contribution to their CO exposure by far comes from their habit, in ambient air the main source is again traffic derived. Ozone is formed by the action of ultraviolet light on oxides of nitrogen and hydrocarbons, so is essentially a pollutant of the summer months in climates such as the UK but may be more perennial in countries where sunlight is present all year round. Ozone levels are generally higher downwind from a city because of the atmospheric chemistry of the formation of ozone combined with the fact that ozone, a very reactive gas, is quickly neutralized by nitric oxide in urban areas.

The health effects of gaseous pollutants have been determined in a number of different ways:

1. By chamber (human challenge) studies
2. By studies of morbidity (*e.g.* symptoms, inhaler use), usually in panels of subjects perceived to be at risk
3. From studies of hospital admissions (*i.e.* routinely collected data)
4. From studies of mortality

1

Chamber studies enable the effects of individual pollutants to be studied alone or in combination with other pollutants on volunteers under strictly controlled conditions. The chief advantage of this type of study is that accurate doses can be delivered and the effects of selected co-factors assessed. However, the volunteers involved in such studies are usually normal subjects or patients with mild asthma who tend to be younger, in contrast to the older subjects who are more likely to be affected by air pollution. Additionally, in chamber studies, the duration of exposure is relatively short compared to outdoor, real-life exposures and consequently it may be difficult to extrapolate findings from these types of studies to the effects that would be seen in the overall population exposed to the outdoor environment. Children are not studied in these types of experiments for ethical reasons, which prevents study of an age group where asthma is very common and in whom the health effects of pollution are often perceived to be significant. However, despite these caveats, chamber studies have, in general, provided very useful information as to the presence or absence of effects of specific pollutants at specific doses and have provided useful insights into the mechanisms of these effects.

Epidemiological studies have been much more informative about health effects both at an individual and population level, studying as they do the real-life situation. The difficulty comes in deciding how large an effect may be and to what specific pollutant or pollutant mix such an effect may be attributable. On a day-to-day basis, exposure to air pollutants may have an immediate effect, either on the same day as a rise in air pollution or perhaps delayed, lagging two, three or more days after a rise. In some situations the cumulative or average exposure over a period of three days or more may be important in determining health outcome. It is even possible that longer lags may be more important for differing health endpoints, an area which is currently being explored.

There is no doubt that there is a range of sensitivities to pollutants across different 'at risk' groups in terms of health effects of air pollution. Patients with pre-existing lung and heart disease appear to be particularly at risk, notably patients with asthma and chronic obstructive pulmonary disease (COPD). More recently, the effects of particulate pollution on patients with coronary heart disease and cerebrovascular disease have been identified, but the role of gaseous pollutants in these two disease categories is not so clear. Asthma is a common condition, affecting around 6% of the total population of the UK. In this condition, the lining of the bronchial tree is inflamed and unduly sensitive to external triggers, such as allergens in those sensitized, viral infections or physical stimuli such as exercise or inhaling cold air. Consequently, these patients are not only important as a risk group for the effects of air pollution but also act as a group where changes in lung function are frequent and measurable when trying to define the presence and size of an effect from an external stimulus. COPD is essentially a disease of cigarette smokers and although, like asthma, it is also an inflammatory condition, on a day-to-day basis these patients show no marked changes in lung function. Patients with either COPD or asthma develop symptoms because of the airway narrowing resulting from the inflammatory process. Where the baseline airway diameter is small, only minor reductions in diameter can produce marked reductions in airflow and hence symptoms.

However, it is at least intuitively logical that, for respiratory diseases, inhalation of polluted air can lead to a deterioration in symptoms.

Coronary heart disease and cerebrovascular disease share a common pathogenesis characterized by the formation of atheroma in the arteries supplying the heart or brain, respectively. In contrast to diseases of the respiratory tract, it is not entirely clear at present how inhalation of air pollutants can lead to vascular health effects, but associations have been shown between ischaemic heart disease deaths and ozone, although the major impacts in this disease area appear to derive from particulate exposure, so further discussion falls outside the remit of this chapter.

It is important to recognize that there may well be interactions between different elements of the pollutant mix in determining health effects. The statistical analysis of time series data (*i.e.* following individuals over long periods of time or considering hospital admissions and mortality over periods of time) will regard each pollutant as a separate entity acting on its own behalf. Because the possible degree of interaction of different pollutants is not known it is impossible to analyse separately for any combined effects. The studies therefore allow for the effects of all other pollutant and non-pollutant factors on that health outcome before determining a residual effect which is then attributed to that pollutant.

2 Quantification of Effect

Quantification of these effects is not easy, but certain guidelines can be used when trying to determine how much of an impact air pollution may have on the public health. A basic concept is that of a threshold. For all the gases considered here (with the probable exception of CO) the assumption has been made that at a population level the effect of the pollutant on health is linearly related and that the relationship passes through zero. Consequently, once the effect size coefficient is known for that pollutant, an estimate of overall effect on the population under consideration can be determined. These quantification estimates will vary from country to country (and almost certainly from area to area within a country) and so we will not consider this further in numerical terms here.

3 Chronic Effects

These discussions apply to the effects of short-term changes in health outcomes which can, in theory, be relatively easily recognized. In contrast, the question of whether long-term exposure over years to particular pollutants or pollutant mixes can lead to long-term health effects as yet remains to be convincingly answered, but may be more important in public health terms. The evidence for such a chronic effect with respect to gaseous air pollutants is scant, whereas there are some data with respect to long-term exposure to particulate pollution which are discussed elsewhere in this volume. Determination of chronic effects is largely dependent upon acquiring data prospectively over a matter of years (longitudinal or cohort studies). Cross-sectional studies where prevalence rates are compared between different areas at the same point in time can contribute to this question to some extent, although they are regarded as being less powerful studies and more likely to be open to uncorrectable confounding. There are no satisfactory longitudinal studies which have considered the effects of gaseous pollutants like the Six Cities Study[1] and the American Cancer Association Study[2] have considered the effects of particulate pollution in this regard. One series of studies of Seventh Day Adventists[3] (an attractive study group as these individuals do not smoke cigarettes, thus removing the major complicating cause of respiratory and heart disease) has suggested that long-term exposure to ozone is associated, in men only, with an increased risk of developing asthma. However, this is an unusual group in an unusual setting and it is not easy to extrapolate these findings to other populations. Consequently, we will only consider the short-term effects in this chapter.

4 Sulfur Dioxide

Controlled Challenge Studies

Normal Subjects.   There is consistent evidence that normal subjects are much less sensitive to the effects of inhaled SO_2 than are patients with asthma. Although one study[4] showed small increases in airways resistance at exposure of 1000 ppb ($2860 \mu g\,m^{-3}$) after a short (ten minute) exposure, other studies have failed to confirm this.[5,6] At exposures of 4 ppm or greater ($11\,440 \mu g\,m^{-3}$), clear effects on airway size have been noted both at rest and with intermittent light exercise.[5,7,8] However, within these averaged group findings a wide range of individual responses can be found, suggesting that there may be individual sub-sets of normal subjects who show a greater response on exposure to this pollutant gas. The clinical significance of these effects is far from clear at present.

There are a number of different factors which may help to explain these variations in response, chief amongst which is the amount of gas entering the lower airways. It is always assumed that SO_2 is a very soluble gas and that if nasal breathing is predominant then doses to the lower respiratory tract will be much reduced because of the nasal trapping of the gas at normal ambient concentrations. A second factor is that some subjects appear to breathe more deeply on exposure to SO_2, thus increasing the dose to the lower respiratory tract. Temperature and humidity can also have a bearing in this regard, in particular cold air which can cause a degree of airway narrowing, although this is only a very small effect in normal individuals, being much more marked in subjects with asthma. However, in challenge studies the effects of these factors should be able to be kept constant between individuals and between exposures within given individuals. It is not

[1] D. W. Dockery, F. E. Speizer, D. O. Stram *et al.*, *Am. Rev. Respir. Dis.*, 1989, **139**, 587.
[2] C. A. Pope, M. J. Thun, M. M. Namboodiri *et al.*, *Am. J. Respir. Crit. Care Med.*, 1995, **151**, 669.
[3] G. L. Euler, D. E. Abbey, J. E. Hodgkin *et al.*, *Arch. Environ. Health*, 1988, **43**, 279.
[4] P. J. Lawther, A. J. Macfarlane, R. E. Waller *et al.*, *Environ. Res.*, 1975, **10**, 355.
[5] N. R. Frank, M. O. Amdur and J. Worcester, *J. Appl. Physiol.*, 1962, **17**, 252.
[6] J. F. Bedi and S. M. Horvath, *JAPCA*, 1989, **39**, 1448.
[7] J. A. Nadel, H. Salem, B. Tamplin *et al.*, *J. Appl. Physiol.*, 1965, **20**, 164.
[8] G. von Nieding, H. M. Wagner, H. Krekeler *et al.*, *Int. Arch. Occup. Environ. Health*, 1979, **43**, 195.

known whether cigarette smoking enhances or inhibits any effects of sulfur dioxide on normal subjects.[9]

Asthmatic Subjects. In patients with asthma, effects on lung function are seen at much lower concentrations. The study by Sheppard *et al.,*[10] while not demonstrating much of an overall effect, showed increases in airways resistance in two very sensitive subjects at an exposure of 100 ppb (286 μg m^{-3}). Other workers, exposing asthmatic subjects to 200 ppb (572 μg m^{-3}), showed small symptom changes but, in a further study, no changes in lung function at 200 ppb exposures associated with heavy exercise.[11] The same group of subjects were exposed to 400 ppb (572 μg m^{-3}) SO$_2$ while undergoing heavy exercise and produced small changes in lung function, but it is not until exposures to levels of around 500 ppb (1430 μg m^{-3}) are employed that there is clear evidence of sulfur dioxide enhancement of exercise-induced airway narrowing.[10,12,13] These responses were seen after exposures of a matter of minutes, whereas, in other studies, longer exposures (up to hours) appeared to be needed to produce an effect. These differences in effect size may be due to differing volunteer characteristics, habituation to the individual's usual air pollutant exposure resulting in tolerance to these levels of laboratory exposure (a recognized phenomenon in studies of ozone challenge) or to methodological differences.

The size of the effect in the challenge studies can be determined by measurement of lung function, the usual measures being those obtained from spirometry, namely the FEV$_1$ (the forced expired volume in 1 second) and the FVC (forced vital capacity). For SO$_2$ the results of exposure on lung function are reasonably consistent across studies, with falls of the order of 50 mL in FEV$_1$ from an approximate start volume of 3 litres for an exposure dose of 200 ppb of SO$_2$. These changes are easily reversible and the size of the effect is small, although if repeated over time these changes may become clinically significant, particularly if the pattern of induced inflammatory change was seen to be relevant to the type of inflammatory change associated with chronic asthma.

Another method of assessing the airway response to a pollutant gas is to measure the change in bronchial responsiveness (bronchial hyper-reactivity) of the individual. This can involve measuring the effect of gas exposure on the response to a non-specific irritant such as methacholine or histamine, a standard method used to characterize the severity of subjects with asthma. Alternatively, the subjects can be exposed to a range of doses of the specific gas and a curve of lung function responses constructed. The study by Horstman *et al.*[14] represents an example of the latter, taking as its main outcome measure the PC$_{100}$ sRaw, the provocative concentration of SO$_2$ causing a 100% increase in specific airways resistance, a sensitive index of flow through larger intrapulmonary airways. Their results show, in a group of asthmatic subjects, a range of responses to sulfur

[9] Advisory Group on the Medical Aspects of Air Pollution Episodes (2nd report), HMSO, London, 1992.
[10] D. Sheppard, A. Saisho, J. A. Nadel *et al.*, *Am. Rev. Respir. Dis.*, 1981, **123**, 486.
[11] W. S. Linn, D. A. Shamoo, C. E. Spier *et al.*, *Environ. Res.*, 1983, **30**, 340.
[12] R. A. Bethel, D. J. Erle, J. Epstein *et al.*, *Am. Rev. Respir. Dis.*, 1983, **128**, 592.
[13] J. R. Balmes, J. M. Fine and D. Sheppard, *Am. Rev. Respir. Dis.*, 1987, **136**, 1117.
[14] D. Horstman, L. J. Roger, H. Kehrl *et al.*, *Toxicol. Ind. Health*, 1986, **2**, 289.

Figure 1 Distribution of individual airway sensitivity to SO_2 as cumulative percentage of 27 subjects. $PC(SO_2)$ is the provocative concentration of SO_2 required to increase airway resistance by 20%

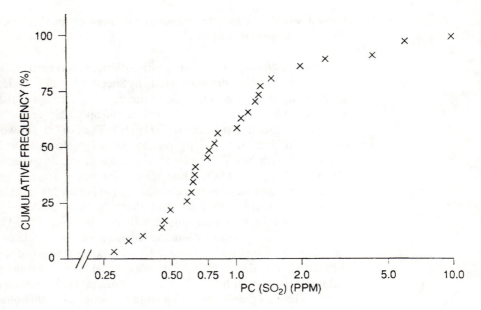

dioxide (Figure 1) with a median PC_{100} sRaw of 750 ppb (2145 $\mu g\, m^{-3}$). However, the usual concentrations of sulfur dioxide seen in ambient air in the UK rarely exceed 120 ppb (343 $\mu g\, m^{-3}$) nowadays, although occasionally, during episodes, levels in excess of 200 ppb have been recorded.

Extrapolating the findings from the chamber studies to the effects on public health is, therefore, somewhat difficult, particularly as those most susceptible to the effects of air pollution (*i.e.* those with more severe disease) are not used in challenge studies. It is likely that these more severely affected individuals have a much lower threshold for developing symptoms or changes in lung function on exposure to air pollution. Consequently, controlled chamber studies can be used to show whether effects in response to a pollutant challenge could occur in a given group of subjects, but extrapolation to all potential members of such a group in real life would be unwise.

Mechanisms. The way in which sulfur dioxide can result in these pathological changes in the airway are likely to be multiple and in some individuals a particular mechanism may be more important than in others. Animal studies[9] show that SO_2 can activate mucosal sensory nerves, leading to airflow obstruction both by central neural reflex and by local axon reflex changes (neurogenic inflammation). Although it is likely that these effects are also true for man, there is no direct work to confirm this. SO_2 may also act by non-neural mechanisms with mucosal damage leading to release of inflammatory mediators, perhaps attracting inflammatory cells, notably neutrophils and eosinophils, to the airway wall.

Morbidity Studies

Studies on morbidity (*i.e.* changes in symptoms and treatment use) are conducted by establishing a cohort (panel) of susceptible individuals and following them prospectively over a period of weeks or months. Over this time the individual will record symptoms twice daily and, in most studies, a measure of lung function such as peak expiratory flow. This approach produces a large amount of data over time at both an individual and a group level. A second way of assessing the response to ambient changes in air pollution is to study individuals during an air pollution episode. However, the quality of information in the latter situation is generally less good as the study is by definition retrospective and opportunistic.

The most important study in the field of panel studies from a methodological point of view was that undertaken by Whittemore and Korn,[15] who clearly spelt out the considerable problems in attributing health effects which were small in size when having to allow for other exposures which could have similar size or greater effects (*e.g.* passive cigarette smoke exposure, exposure to animals in sensitized asthmatic subjects). They also stressed the importance of allowing for other pollutants when trying to assess the effects of a given pollutant using multiple regression analysis to try and pick out a signal from the noise. Many studies, however (and this applies to epidemiological studies such as those assessing hospital admissions and mortality), use single pollutant models in their analysis rather than two or multi-pollutant models. However, providing the potential shortcomings of the methodological approaches are recognized, these studies do inform on the problems of determining these generally small health effects. This method of analysis has been followed by many subsequent researchers who have also, with variable success, attempted to deal with confounders and co-exposures in panel studies. Most of these studies have, however, been undertaken in children, either specific groups of subjects with asthma or school summer camp studies where a variable number have asthma. A significant problem comes when recruiting subjects for such studies and having to define whether a child with respiratory symptoms should be regarded as having asthma or another diagnosis, such as episodic bronchitis. In some studies, this problem has been circumvented by classifying subjects into those with and without symptoms.[16] Studies in adults are infrequent.

The size of any effect needs to be quantified in relation to the air pollutant exposure and estimates have been expressed as a change per unit change in pollutant, for a larger change in level (*e.g.* per 10 μg m^{-3} rise in particles) or for a change over the interquartile range over which the levels of that pollutant were measured during the running of that particular study. Unfortunately, these different approaches mean that it is not always possible to undertake comparisons between studies. Because most of the studies from the US and Canada have been undertaken in the summer, the main pollutants assessed have been particles, ozone and aerosol strong acid (the ozone studies being discussed below), SO$_2$ being regarded until lately by North American workers as an unimportant pollutant for their countries, largely on the basis of low measured ambient levels.

[15] A. S. Whittemore and E. L. Korn, *Am. J. Public Health*, 1980, **70**, 687.
[16] S. Vedal, M. B. Schenker, A. Munoz *et al.*, *Am. J. Public Health*, 1987, **77**, 694.

Figure 2 Daily change in symptom score in patients with chronic bronchitis over time in relation to levels of SO_2 and black smoke

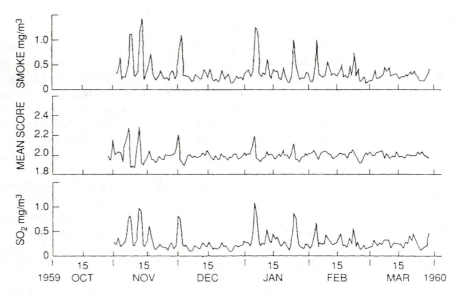

Where SO_2 has been considered by US workers, health effects were usually absent (summarized in ref. 17). In one study of symptomatic and asymptomatic children in Pennsylvania,[18] reductions in peak flow of the order of 0.9 L min^{-1} were identified for an increase in SO_2 of $15 \,\mu\text{g m}^{-3}$ (5 ppb). The baseline value of peak flow in this group was of the order of 300 L min^{-1} and such a fall is clinically insignificant on a day-to-day or one-off basis. European panel studies have shown effects relating to SO_2 in three of four studies,[19-22] but with effect sizes of similar degree. The Manchester study[22] also suggested that those with more severe asthma (as measured by bronchial responsiveness) were more likely to show an effect of SO_2, although particulate levels were not measured in this study and the SO_2 effect could be due to particles alone.

In the early 1960s, when SO_2 levels were around $300 \,\mu\text{g m}^{-3}$ (105 ppb) with regular peaks to $1000 \,\mu\text{g m}^{-3}$ (350 ppb), a series of panel studies of patients with chronic bronchitis (now defined as chronic obstructive pulmonary disease, or COPD) showed quite marked day-to-day correlations between symptoms and levels of both smoke and sulfur dioxide (Figure 2).[23] The symptom scoring system was fairly simple but proved robust. When a similar group of patients were studied some 10 years later when sulfur dioxide levels had dropped to around $200 \,\mu\text{g m}^{-3}$ (70 ppb) [with only rare excursions above $500 \,\mu\text{g m}^{-3}$ (175 ppb)], the day-to-day changes in symptoms were not perceptibly correlated. This suggested

[17] Committee on the Medical Effects of Air Pollutants, *Asthma and Outdoor Air Pollution*, HMSO, London, 1995.

[18] L. M. Neas, D. W. Dockery, J. D. Spengler *et al.*, *Am. Rev. Respir. Dis.*, 1992, **145**, A429.

[19] W. Roemer, G. Hoek and B. Bruenkreef, *Am. Rev. Respir. Dis.*, 1993, **147**, 118.

[20] A. Peters, U. Beyer, C. Spix *et al.*, *Am. J. Resp. Crit. Care Med.*, 1994, **149**, A662.

[21] S. M. Walters, J. G. Ayres, G. Archer *et al.*, *Am. J. Respir. Crit. Care Med.*, 1994, **149**, A661.

[22] B. G. Higgins, H. C. Francis, C. J. Warburton *et al.*, *Thorax*, 1993, **48**, 417.

[23] R. E. Waller, *J. R. Coll. Physicians London*, 1971, **5**, 362.

that either effects were no longer present at these background levels or, more likely, that they were not detectable using this fairly crude methodology. However, the apparent absence of measurable changes suggests that on a day-to-day basis such levels of SO_2 will not cause problems in the majority of subjects with COPD, although the question of the effects of exposures at this level over longer time periods remains.

Air pollution events featuring sulfur dioxide are now rare in the western world, although where there is considerable burning of fossil fuels these events may be more frequent. The London fog incident of 1952[24] was associated with huge exposures to SO_2 (3 ppm) and resulted in very substantial increases in mortality and morbidity, although discussion still takes place as to whether the particles, sulfur dioxide or the acidity of the pollutant mix was the most responsible component. A study from Holland of an event occurring in the late 1960s[25] showed a clear and significant reduction in lung function in normal subjects during and following an air pollution episode when SO_2 levels rose to 300 μg m^{-3} (105 ppb). The degree of loss was 150 mL in FEV_1, an approximate 5% loss. More importantly, this loss of lung function was still present two weeks later, although it returned to normal thereafter. Later studies from the same country[26,27] showed similar effects also in normal subjects. More recent air pollution events in the UK have been associated with relatively modest increases in sulfur dioxide. The London event of 1991[28] showed increases in sulfur dioxide to around 125 ppb, although the Birmingham event of 1992 was associated with levels of around 250 ppb.[29] The London event was associated with a 10% increase in all cause mortality, but the only measurable health effect at the primary care level was a small increase in consultations for sore throat. Particles appeared to be the most important causal factor, sulfur dioxide not having an effect once particles were accounted for. The Birmingham study[29] considered two panels of subjects with mild and with severe asthma and was followed using daily diary cards over the period of the pollution episode. There were no discernible effects on the patients with mild asthma but those with severe asthma showed a significant fall in lung function during the period despite increasing their inhaled and oral therapy. It would appear that pollution events where SO_2 was increased, at an individual level, may have limited effects except in those with more severe disease, although there still appears to be some discrepancy between studies from different countries. This again flags up the difficulty in extrapolating the findings from air pollution studies from country to country, at least in quantitative terms.

In 1990 the WHO determined that for sulfur dioxide levels of around 200 μg m^{-3} (70 ppb), small transient reductions in lung function could be seen in children and adults that could last for as much as two to four weeks, but the magnitude of the effect was small at around 2–4%. However, because airborne levels of sulfur dioxide and particulate matter move so clearly together over time

[24] Ministry of Health, HMSO, London, 1954.
[25] R. van der Lende, C. Huygen, E. J. Jansen-Koster *et al.*, *Bull. Physiopathol. Respir.*, 1975, **11**, 31.
[26] B. Brunekreef, M. Lumers, G. Hoek *et al.*, *JAPCA*, 1989, **36**, 1223.
[27] G. Hoek, B. Brunekreef, P. Hofschreuder *et al.*, *Toxicol. Ind. Health*, 1990, **6**, 189.
[28] H. R. Anderson, E. S. Limb, J. M. Bland *et al.*, *Thorax*, 1995, **50**, 1188.
[29] S. M. Walters, J. Miles, J. G. Ayres *et al.*, *Thorax*, 1994, **48**, 1063.

and space, the WHO found it difficult to apportion these effects entirely to the gas or the particle phase. Their interpretation of the research available up to that time led them to state that if levels of SO_2 reached $250 \mu g m^{-3}$ (87 ppb) there was a measurable increase in respiratory morbidity amongst susceptible adults with COPD and perhaps also in children, these figures being more marked when levels reached around $400 \mu g m^{-3}$ (140 ppb).[30]

Hospital Admissions

The re-emergence of SO_2 as an important air pollutant has largely occurred as a result of the series of multicentre studies called the APHEA studies (Air Pollution and Health, a European Approach). The cities involved ranged from Helsinki in the north to Barcelona in the south. Although there were some methodological differences across the cities involved in the study in terms of data acquisition, the statistical approach was uniform and as robust as could be for the data. In the event, the consistency and coherence of the results suggest that the findings are not only valid but also very important, at least as regards determination of effects in the European setting. The data are expressed as a relative risk for a standard increase in pollutant and was taken as $50 \mu g m^{-3}$ for all pollutants including SO_2.[31] For all respiratory admissions the overall effect for all cities involved for this increment in SO_2 was 1.009 (95% confidence interval 0.992–1.025) for individuals between the ages of 15 and 64 increasing to 1.020 (1.005–1.046) for those over the age of 65.[32] To put this into a clinical context, these admissions relate essentially to patients with COPD or asthma, the former making by far the largest contribution. For COPD, such patients have pre-existing disease, whereas in asthma the patients affected may be a mix of those with moderate to severe disease through to those with milder asthma who, on the day of the increase, may not have had adequate self-treatment available. A similar effect was seen during the outbreak of asthma attacks seen in the UK in 1994 during a thunderstorm.[33] An increase in admissions of 2% on a day when SO_2 increases by $50 \mu g m^{-3}$ represents a very small risk at an individual level but can result in a significant effect in public health terms, assuming that the relationship between admission and pollutant level is linear through zero, which is generally accepted to be the case for SO_2. However, there was considerable heterogeneity of results for SO_2 in this study across the cities involved, the more homogeneous results being seen for ozone (see below). These findings are in contrast to the study from Birmingham,[34] which showed more of an effect from black smoke than SO_2, although a small effect of the gaseous component of the pollutant mix was seen in that study. Further breakdown of the APHEA studies for asthma and for COPD perhaps add a little clarity to the picture. For COPD, the relative risk for six cities was 1.022 for SO_2 (for a $50 \mu g m^{-3}$ rise) although the confidence intervals around this estimate embraced unity. However, for Paris, Milan and Barcelona the risk was

[30] World Health Organization, European Series No. 43, Copenhagen, 1992.
[31] C. Spix, H. R. Anderson, J. Schwartz *et al.*, *Arch. Environ. Health*, 1998, **53**, 54.
[32] R. Newson, D. Strachan, E. Archibald *et al.*, *Thorax*, 1997, **52**, 680.
[33] S. Walters, R. K. Griffiths and J. G. Ayres, *Thorax*, 1994, **49**, 133.
[34] H. R. Anderson, C. Spix, S. Medina *et al.*, *Eur. Respir. J.*, 1997, **10**, 1064.

Figure 3 Estimated individual city and pooled relative risks of mortality associated with an increase in SO$_2$ level of 50 μg m^{-3}. Numbers in brackets are median values for SO$_2$ and the size of the point representing each relative risk is inversely proportional to its variance

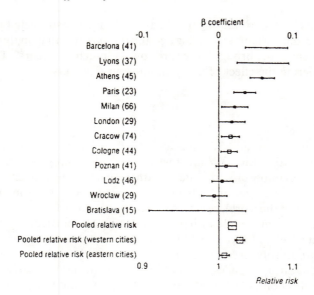

higher and statistically significant.[35] For asthma the effect size was smaller still and did not achieve statistical significance.[36] This may be a feature of the smaller numbers of asthma admissions compared to those for COPD.

Mortality

Data from the severe smogs of the 1950s in the UK and from other parts of the world enabled the WHO[37] to consider that increases in daily deaths become discernible when 24 hour average concentrations of sulfur dioxide exceed about 175 ppb SO$_2$ (500 μg m^{-3}) with equivalent levels of black smoke. It was always deemed a matter of concern that extrapolating from these historical data to the 1990s was unwise, but recent studies by the APHEA group have shown that sulfur dioxide consistently emerges as a factor for mortality.[38] In western European cities, a 3% (95% CIs 2–4%) increase in daily mortality (all causes) was found for a 50 μg m^{-3} rise in either SO$_2$ or black smoke. However, in eastern European cities the effects were much smaller (Figure 3) with an increase in all cause mortality of just 0.8% (95% CIs −0.1 to +2.4%) for the same rise in SO$_2$. These effects were seen whether assessing a same-day effect or a cumulative effect over 2–4 days.

It is important to realize that exposure to air with higher levels of air pollutants will contribute to death by bringing forward the time of death. Patients with COPD, for instance, who die on days of higher pollution will already have severe pre-existing disease. Consequently it is likely that the time of death may have been brought forward by only a small amount, perhaps days. On the other hand,

[35] J. Sunyer, C. Spix, P. Quenel *et al.*, *Thorax*, 1997, **52**, 760.
[36] World Health Organization, Environmental Health Criteria No. 8, Geneva, 1979.
[37] K. Katsouyanni, G. Touloumi, C. Spix *et al.*, *Br. Med. J.*, 1997, **314**, 1658.
[38] Advisory Group on Medical Aspects of Air Pollution Episodes (3rd report), HMSO, London, 1993.

patients with coronary heart disease may have unsuspected pre-existing severe disease and perhaps at a young age and it is possible that, in this group of patients, death may be brought forward by as much as years. This applies when considering the effect of any pollutant on mortality.

Summary

In challenge studies, SO_2 has a clear effect on airway function, particularly in patients with asthma. While it was believed for many years that this gas was no longer an important pollutant with respect to health effects because of the marked reductions in ambient levels produced by Clean Air Act legislation throughout the world, recent evidence, particularly from Europe, has shown that SO_2 impacts on people with pre-existing disease, particularly respiratory disease. Despite ambient levels of SO_2 which are very low compared to those seen in Western European countries in the 1950s and 1960s, these effects can be identified and range from changes in symptoms to hospital admissions and mortality. From the health point of view, attention needs to be paid to the control of a pollutant which is not vehicle generated.

5 Nitrogen Dioxide

Controlled Challenge Studies

Exposure of normal subjects or subjects with pre-existing lung conditions to nitrogen dioxide in the challenge situation has generally resulted in no changes in either lung function or symptoms at low to moderate levels of exposure, although very high levels (in excess of 2 ppm) can result in some changes in certain, presumably susceptible, individuals.

Normal Subjects. In normal subjects a wide range of studies with exposures of 1000 ppb or less over periods ranging from 20 minutes to two hours had no effect on any lung function measurement or index of bronchial responsiveness.[39] At a range of exposures between 1.5 ppm for three hours up to 7.5 ppm for two hours, non-specific bronchial responsiveness to methacholine was increased to a small extent.[40-42] These effects do not appear to be affected by increasing age, but no studies have been undertaken in children to address the lower end of the age spectrum.

Patients with COPD. In this group of patients the findings are more variable. In some studies, exposure to high levels (1.5 ppm) of NO_2 ($2820 \mu g \, m^{-3}$) showed an increase, albeit small, in airways resistance,[43] while a dose–response study with

[39] M. Beil and W. T. Ulmer, *Int. Arch. Occup. Environ. Health*, 1976, **38**, 31.
[40] V. Mohsenin, *Arch. Environ. Health*, 1988, **43**, 242.
[41] M. W. Frampton, P. E. Morrow, C. Cox *et al.*, *Am. Rev. Respir. Dis.*, 1991, **143**, 522.
[42] W. S. Linn, J. C. Solomon, S. C. Tree *et al.*, *Arch Environ. Health*, 1985, **40**, 234.
[43] G. von Nieding, M. Wagner, H. Krekeler *et al.*, *Int. Arch. Arbeitsmed.*, 1971, **27**, 338.

doses ranging from 0.5 to 2 ppm over an hour with 30 minutes exercise showed no effect; similar findings have been found in a number of other studies.[38]

Asthmatic Subjects.   The findings in asthmatic subjects are not too dissimilar from those found in normal subjects. Apart from one study[45] of exposure to 100 ppb of NO_2 for one hour, which appeared to increase methacholine responsiveness, the majority of studies up to and including exposure to 4 ppm have shown no change in lung function and no change in bronchial responsiveness.[38] However, there are a few studies[46-50] where some change in bronchial responsiveness did occur with exposures between around 250 and 500 ppb for durations ranging between 30 minutes and three hours. More recently, nitrogen dioxide exposure at a level of 400 ppb has been shown to enhance the subsequent response to allergen challenge,[51] suggesting that NO_2 on its own probably has very limited effect, even in patients with asthma, when acting directly, but may act as a priming agent to other co-exposures such as allergen challenge, with sensitized subjects showing an enhanced (+ 5%) bronchoconstrictor response to allergen challenge following NO_2 compared to air.

Mechanisms.   The mechanism of any priming effect is not entirely clear at present. There is some evidence that eosinophils are recruited and activated on exposure to NO_2[52] and some suggestion of alteration in inflammatory cell sub-sets in bronchial alveolar lavage fluid following NO_2 exposure. Further work is needed to explore this subtle potential effect of exposure to this pollutant gas.

Morbidity

The findings for morbidity are inconsistent and this has led to some lack of conviction that any effect shown of NO_2 on symptoms or treatment use is in fact due to residual confounding rather than a genuine effect. It is felt by many that the most likely explanation of this dissimilarity in results is that day-to-day changes in morbidity indicators associated with NO_2 are in fact causally related to another pollutant which moves closely with nitrogen dioxide in terms of ambient levels. Perhaps the most likely is another index of ultrafine particles such as $PM_{1.0}$ or particle numbers.[53] However, many studies have found associations between NO_x linked pollution and health effects and at levels well below the WHO guidelines.

It is probably inappropriate, therefore, to regard NO_2 as a gas which we can be confident about identifying as a cause of day-to-day changes in morbidity on

[44] W. S. Linn, D. A. Shamoo, E. L. Avol *et al.*, *Arch. Environ. Health*, 1985, **40**, 313.
[45] J. Orehek, J. P. Massari, P. Gayrard *et al.*, *J. Clin. Invest.*, 1976, **57**, 301.
[46] G. Bylin, T. Lindvall, T. Rehn *et al.*, *Eur. J. Respir. Dis.*, 1985, **66**, 205.
[47] M. A. Bauer, M. J. Utell, P. E. Morrow *et al.*, *Am. Rev. Respir. Dis.*, 1986, **134**, 1203.
[48] V. Mohsenin, *Toxicol. Environ. Health*, 1987, **22**, 371.
[49] G. Bylin, G. Hedenstierna, T. Lindvall *et al.*, *Eur. Respir. Dis.*, 1988, **1**, 606.
[50] R. Jorres and H. Magnussen, *Eur. Respir. J.*, 1990, **3**, 132.
[51] W. S. Tunnicliffe, P. S. Burge and J. G. Ayres, *Lancet*, 1994, **344**, 1733.
[52] J. L. Devalia, R. J. Sapsford, D. R. Cundell *et al.*, *Eur. Respir. J.*, 1993, **6**, 1308.
[53] J. G. Ayres, *Eur. Respir. Rev.*, 1998, **8**, in press.

present evidence. However, the longer-term effects of nitrogen dioxide exposure, particularly when one bears in mind that oxides of nitrogen are major products of gas combustion in kitchens and levels are also elevated in rooms with a poorly serviced gas fire, may be more important in determining chronic health effects.

There have been many studies going back a number of years[54,55] showing that respiratory symptoms in population studies were more noticeable in homes with gas fired kitchens. More recently, one of the UK centres of the European Respiratory Health Study[56] showed that in women (although not in men) respiratory symptoms were greater in those who had gas fired kitchens compared to those with electrically powered homes. There was a tendency for the symptoms to be worse in those who were atopic (*i.e.* sensitized to allergens), although this did not quite achieve statistical significance. More extensive analysis of the full European project has, however, shown that this finding is not uniform across all the geographical areas studied.

Cross-sectional studies of disease prevalence can provide information on the effects in populations differentially exposed to air pollutants. A good example is the series of studies of re-unified Germany, where the population of old Eastern Germany was previously exposed to high levels of SO_2 and black smoke compared to the higher exposures in Western Germany to NO_x.[57] The association between a higher prevalence of productive cough (bronchitis) in children with SO_2 in Eastern Germany and of NO_2 in Western Germany with asthma and hayfever initially led to the, probably incorrect, assumption that NO_2 was a cause of asthma and other allergic diseases. Studies over time, when SO_2 and black smoke levels have declined considerably in old Eastern Germany, have shown a fall in the prevalence of bronchitis in children but no increase in asthma or hayfever. Apart from being fairly certain that the older, sulfurous, air pollution did lead to mucus hypersecretion, these figures do not yet help us in determining whether NO_2 has a role to play in the aetiology of allergic diseases.

Hospital Admissions

The findings from a range of studies of hospital admissions or attendances to Accident & Emergency Departments are similar to those from the morbidity studies. The majority show no effect at all, while others do find associations for a range of conditions with day-to-day changes in NO_2 levels.[38] A study from Vancouver[58] did find a correlation between A&E attendances for respiratory disease in subjects over the age of 60, but NO_2 levels were strongly correlated with other pollutants. A European study[59] showed that for a $10\,\mu g\,m^{-3}$ rise in NO_2, hospital attendances for croup increased by around 4%, but the association was as strong for particles and it is difficult, given these findings, to separate out either as the more important causal pollutant.

[54] R. J. W. Melia, C. du V. Florey, D. G. Altman *et al., Br. Med. J.*, 1972, **2**, 149.

[55] R. J. W. Melia, C. du V. Florey, R. W. Morris *et al., Int. J. Epidemiol.*, 1982, **11**, 164.

[56] D. J. Jarvis, S. Chinn, C. Luczynska *et al., Lancet*, 1996, **347**, 426.

[57] E. von Mutius, Ch. Fritzsch, S. K. Weiland *et al., Br. Med. J.*, 1992, **305**, 1395.

[58] D. V. Bates, M. Baker-Anderson and R. Sizto, *Environ. Res.*, 1990, **51**, 51.

[59] K. Katsouyanni, A. Karakatsani, I. Messari *et al., J. Epidemiol. Community Health*, 1990, **44**, 321.

Mortality

Where NO_2 has been associated with changes in mortality, associations have also been shown with other pollutants so that disentanglement of the true effects pollutant by pollutant is not possible.[38] The APHEA studies[37] have not been helpful in this regard as insufficient NO_2 data were available measured over suitable time sampling frames. However, the current set of epidemiological studies (APHEA 2) will hopefully be able to consider the role of NO_2 in this respect, although the likelihood is that any effect possibly due to NO_2 will be inseparable from the effects due to other pollutants.

Summary

In summary, while it is attractive to view NO_2, an oxidant pollutant, as a cause of health effects, the evidence from challenge and epidemiological studies do not show significant health effects. It is likely, however, that longer-term exposures, especially in the indoor setting, may have an impact in terms of chronic effects, maybe by a permissive mode of action allowing other stimuli to have a greater effect on the individual.

6   Carbon Monoxide

Mechanisms

Unlike the other gaseous pollutants covered in this chapter, the mechanism whereby CO exerts an effect on human health is well understood, at least at the level of its ability to bind to haemoglobin. More recently, to a certain extent as a result of studies of CO in the outdoor environment and its health effects, suggestions have been made that this gas may also act in a different way, although a satisfactory alternative mechanism has yet to be supplied.

In toxic doses, such as might be encountered in severe poisoning, the first sign is loss of consciousness. If death does not then ensue, some patients are left with cerebral damage of variable degree, with a wide range of symptoms and clinical signs. These effects are due to the formation of carboxyhaemoglobin (COHb) in red blood cells, at levels which can be measured in blood or, indirectly, measured as CO in exhaled breath. In normal, unexposed subjects, COHb levels are 1% or below. Cigarette smokers may have levels ranging from 4% to 15%, depending on the number of cigarettes smoked. Consequently, many smokers will contribute to ambient levels, their exhaled breath CO concentrations exceeding that of ambient air.

Challenge Studies

At COHb concentrations of 2.5–4.0%, effects have been seen on maximal exercise duration in fit healthy men;[60] at levels slightly higher, patients with angina find that episodes of chest pain occur earlier during exertion than when their COHb is

[60] J. A. M. Turner and M. W. Nicol, *Respir. Med.*, 1993, **87**, 427.

in the normal range. It has thus been believed that ambient levels of CO producing COHb levels below 2.5% are unlikely to exert any deleterious effects even in patients with coronary heart disease. This level would be achieved by breathing 10 ppm CO for 8 hours, 25 ppm for 1 hour, 50 ppm for 30 minutes or 87 ppm for 15 minutes.

Morbidity Studies

Recently, reports from North America and London suggest that hospital admissions for heart failure are related to ambient levels of CO, increasing by around 23% for a 10 ppm rise in CO.[61] Studies from Athens and Los Angeles also show an effect on all cause mortality, with effect sizes ranging from 4 to 11.5% for a 10 ppm rise in CO.[61] These findings are difficult to interpret, given the current knowledge of mechanisms. Either the result is spurious, being due to residual confounding, or the effect is real and due to the fact that those patients with severe congestive cardiac failure are very sensitive to small changes in COHb, or there is an alternative mechanism whereby CO is acting on these patients. Given that many patients with cardiac failure are smokers of cigarettes, and the dose delivered to the lung in these subjects from their habit hugely exceeds ambient exposures, these findings might imply that unrecognized confounders may explain these results rather than it being a specific effect in susceptible groups, but targeted studies stratified by disease severity would help to address this issue.

Summary

Ambient CO exposure is generally low in urban areas compared to exposure to CO through actively or passively inhaled cigarette smoke, and although some epidemiological evidence would suggest that, perhaps in certain susceptible individuals with heart disease, ambient levels can be associated with health effects, the mechanisms for these findings are far from clear. It would be prudent, however, to consider further these possible health effects, making careful allowance for co-exposures, to determine whether these effects are real or spurious.

7 Ozone

Controlled Challenge Studies

One of the main differences between ozone—a highly reactive, oxidant gas as far as the airways are concerned—and sulfur dioxide with respect to challenge data is that ozone appears to affect both asthmatics and normal subjects equally, whereas sulfur dioxide is much more likely to cause airway narrowing in patients with asthma.[62] Inflammatory changes have been shown histologically in the bronchial mucosa in animals at ozone concentrations as low as 80 ppb. Other animal work suggests that, at much higher exposures, irreversible pulmonary

---

[61] Committee on the Medical Effects of Air Pollutants, *Quantification of the Effects of Air Pollution in the UK*, The Stationery Office, London, 1998.

[62] Advisory Group on the Medical Aspects of Air Pollution Episodes (1st report), HMSO, London, 1991.

fibrosis or emphysema can occur.[63] In the human setting, however, most studies have involved normal or mild asthmatic subjects undergoing fairly strenuous exercise in an ozone chamber for several hours, trying to reproduce the sort of exposure that might occur during periods of persistently high ozone levels in the normal environment.

Normal Subjects. At doses varying between 80 ppb and 600 ppb for periods ranging from one hour to six hours, there is evidence of increases in specific airways resistance at the higher end of the range of exposures and also enhancement of non-specific responsiveness to both histamine and methacholine.[64-66] However, there is wide variation in the response to ozone within the subjects constituting the normal groups in these studies. For instance, in the study by Horstman *et al.*[65] the fall in FEV_1 after exposure to 120 ppb of ozone for over six hours ranged from zero in about one third of the subjects to 37%. Whether smokers are more susceptible than non-smokers to higher concentrations of ozone remains unclear.[67]

Asthmatic Subjects. Studies in asthma have looked at similar types of exposures both in terms of dose and duration and the co-existence of exercises, the normal studies. However, although there are changes in airways resistance,[68,69] these results are generally small and appear to be of the same order as normal subjects. One interesting finding is that repeated exposures on consecutive days appear to be related to the amelioration of effects on symptoms in patients with asthma, suggesting a role for the development of tolerance to ozone in those repeatedly exposed.

More importantly, ozone appears to act as a permissive agent by enhancing the response to another co-exposure. The best example of this is a response to allergen, particularly grass pollen, as grass pollen levels in ambient air in most temperate countries will be elevated during times when ozone events are more likely to occur. The best study characterizing this response[70] covered 24 asthmatic subjects, 12 with allergic rhinitis and 10 controls, exposing them to ozone (250 ppb) or filtered air and assessing the effect on subsequent grass pollen challenge. In the asthmatic subjects, the amount of allergen required to reduce FEV_1 by 20% was lower by a factor of 1.74 after ozone compared to air challenge, with similar changes in the rhinitis group. This is similar in extent to the changes found for NO_2 and house dust mite allergen challenge[51] and provides a coherence of evidence for this more subtle effect of gaseous air pollutants. The use of grass pollen challenge with ozone is very relevant as in many temperate climates the ozone levels are only significantly increased during the summer

[63] B. E. Barry, F. J. Miller and J. D. Crapo, *Lab. Invest.*, 1985, **53**, 692.

[64] L. J. Folinsbee and M. H. Mazucha, in *Atmospheric Ozone Research and its Policy Implications*, eds. T. Schneider, S. D. Lee, G. J. R. Wolters and L. D. Grant, Elsevier, Amsterdam, 1989.

[65] D. H. Horstman, L. J. Folinsbee, P. J. Ives *et al.*, *Am. Rev. Respir. Dis.*, 1990, **142**, 1158.

[66] M. J. Holtzman, J. H. Cunningham, J. R. Sheller *et al.*, *Am. Rev. Respir. Dis.*, 1979, **120**, 1059.

[67] M. Hazucha, F. Silverman, C. Parent *et al.*, *Arch. Environ. Health*, 1973, **27**, 183.

[68] J. W. Kreit, K. B. Gross, T. B. Moore *et al.*, *J. Appl. Physiol.*, 1989, **66**, 217.

[69] J. Q. Koenig, D. S. Covert, Q. S. Hanley *et al.*, *Am. Rev. Respir. Dis.*, 1990, **141**, 377.

[70] R. Torres, D. Nowak and H. Magnussen, *Am. J. Respir. Crit. Care Med.*, 1996, **153**, 56.

months when grass pollen counts or grass allergen levels in the air are increased. The concentration of ozone used in this study is high compared to the usual ambient ozone levels (even during high ozone spells) found in such areas of the world, but this study has at least shown that such an effect does occur. More recently, studies employing lower dose allergen challenge over longer time exposures, producing more realistic environmental co-exposures, have been undertaken which suggest that potentiation of allergen response by gaseous pollutants can still be identified (V. Strand, personal communication).

Mechanisms.   Being an oxidizing gas, ozone is known to enhance inflammation in both the proximal and distal intrapulmonary airways, but the exact mechanism by which this occurs is still undetermined. There is evidence that ozone induces airway neutrophilia in normal subjects at doses of 200[71] to 300 ppb[72] for one hour and these changes are more likely to be seen in asthmatic compared to normal subjects.[71] In addition, a range of pro-inflammatory cytokines and markers of inflammation (*e.g.* Il8, GM-CSF and myeloperoxidase) are released into the airways during ozone challenge, again being more marked and more widespread in asthmatic subjects.[71] More recently, ozone exposure at 160 ppb has been shown to induce eosinophil infiltration in the airways of asthmatic subjects, the eosinophil being the crucial inflammatory cell in asthma.[73] A single exposure to ozone is thus able to induce or enhance inflammation in the airways of both normal and asthmatic subjects but more so in the latter where inflammation already exists. It is not known what occurs in repeated or chronic exposure, although repair systems will undoubtedly come into play to reduce or modify the response.

Morbidity Studies

Most of the studies of ozone on day-to-day changes in symptoms and treatment use have come from North America, largely from children attending summer camps. These often have a preponderance of asthmatic subjects but by no means always. Again, there is a range of responses both in terms of lung function and in symptoms. Lung function has shown changes over a range of 0.1–1.1 mL ppb^{-1} in forced vital capacity and approximately the same deficit for the forced expiratory volume in one second (FEV_1). For peak flow the changes vary between 3 and 7 mL s^{-1} ppb^{-1}.[62] These are very small changes and in general are unlikely to be noticed at a clinical level. These mean data, however, do cover a wide range of response. For instance, Figure 4[74] shows that the median decline in forced vital capacity was 5 mL ppb^{-1} ozone (range −20 to +25 mL ppb^{-1}). This suggests that there may be susceptible sub-populations who are more responsive to ozone and tends to match the findings from the challenge studies.

The data on those panel studies where adults have been the subjects are very

[71]   C. Scannell, L. Chen, R. M. Aris *et al.*, *Am. J. Respir. Crit. Care Med.*, 1996, **154**, 24.
[72]   E. S. Schelegle, A. D. Siefkin and R. J. McDonald, *Am. Rev. Respir. Dis.*, 1991, **143**, 1353.
[73]   D. B. Peden, B. Boehlecke, D. Horstman *et al.*, *J. Allergy Clin. Immunol.*, 1997, **100**, 802.
[74]   P. L. Kinney, J. H. Ware, J. D. Spengler *et al.*, *Am. Rev. Respir. Dis.*, 1989, **139**, 56.

Figure 4 Distribution of
changes in FVC
(mL ppb^{-1} ozone change)
in children

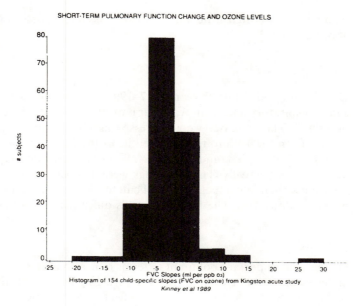

SHORT-TERM PULMONARY FUNCTION CHANGE AND OZONE LEVELS

Histogram of 154 child-specific slopes (FVC on ozone) from Kingston acute study
Kinney et al 1989

limited. One UK study showed no effect of ozone on either symptoms or lung function.[21]

As far as symptoms are concerned, some studies have demonstrated more marked changes in symptoms than in lung function, some showing a dose–response relationship. A study from Los Angeles, where ozone exposure is high and persistent, showed that symptoms of chest discomfort and cough increased by around 25% when levels exceeded 300 ppb, but that above 400 ppb, chest discomfort increased by more than twofold and cough by 77%.[75] However, even at lower levels of exposure, relationships can be found between symptoms such as breathlessness and ozone exposure.[76]

Hospital Admissions

The first indication that ozone might have an impact on hospital admissions came from studies in Canada,[58,77] where it was more the acidic component of the summer pollutants that appeared to be correlated to hospital admissions for asthma rather than ozone itself. Further studies in the UK (London) failed to show an effect of ozone on hospital admissions for cardiovascular diseases,[78] although other studies (see ref. 79) have shown small but significant positive associations. For example, London data from different years suggested a relative risk for admission of 1.04 for a 25 ppb increase in ozone concentration. A modelling process using this coefficient has suggested that, in the UK, if using a 50 ppb threshold, ozone was involved in about 0.25% of all respiratory

[75] J. Schwartz and S. Zeger, *Am. Rev. Respir. Dis.*, 1990, **141**, 62.
[76] B. Ostro, M. Lipsett, J. Mann *et al.*, *Am. J. Respir. Crit. Care Med.*, 1994, **149**, A658.
[77] D. V. Bates and R. Sizto, *Environ. Res.*, 1987, **43**, 317.
[78] J. D. Poloniecki, R. W. Atkinson, A. Ponce de Leon *et al.*, *Occup. Environ. Med.*, 1997, **54**, 535.

admissions, although if no threshold was assumed this estimate increased by about 20-fold.[79]

Mortality

Data from London for the years 1987–1992 have shown an effect of ozone on all cause, respiratory and cardiovascular mortality, particularly during the warmer months.[80] The increases were expressed as the % change in mortality for a change from the 10th to the 90th centile of the measured levels, which yielded values of 3.5%, 3.6% and 5.4% for all cause, cardiovascular and respiratory mortality, respectively. These findings were independent of the effects of other pollutants but might appear to be difficult to square with the lack of effect seen for hospital admissions. The reasons for this difference are not as yet understood.

Summary

Ozone is an important pollutant in terms of health effects, with clear impacts at all levels of morbidity and an effect on bringing forward death. It is a seasonal pollutant and thus has a much greater effect in the summer months in temperate climes but is a year-round pollutant in areas with long hours of sunshine. It should be regarded as a non-threshold pollutant so that health effects can in theory be seen at all levels to zero.

8   Overall Conclusions

The most important gaseous pollutants with respect to human health are SO_2 and ozone, at least in terms of acute effects. Both seem to affect patients with respiratory disease, affecting asthmatic subjects at the level of symptoms, although with less of an effect on attacks severe enough to result in hospital admissions, and patients with COPD, particularly those with severe disease, resulting in hospital admission and advancement of death. The evidence for NO_2 exerting an effect on health on a day-to-day basis is weak but it may play a role, particularly when considering indoor exposures, on chronic respiratory disease states. If these health effects are to be ameliorated, attention needs to be paid as much to sources of SO_2 (power stations) as of ozone (vehicle emissions), while for NO_2 the quality of indoor air needs to be addressed.

[79] J. R. Stedman, H. R. Anderson, R. W. Atkinson *et al.*, *Thorax*, 1997, **52**, 958.
[80] H. R. Anderson, A. Ponce de Leon, J. M. Bland *et al.*, *Br. Med. J.*, 1996, **312**, 665.

The Mechanism of Lung Injury Caused by PM$_{10}$

KEN DONALDSON AND WILLIAM MacNEE

1 Adverse Health Effects of PM$_{10}$

This article addresses the mechanisms of the adverse health effects of PM$_{10}$, *i.e.* the mass of particulate air pollution collected by a convention that has 50% efficiency for particles with an aerodynamic diameter of 10 μm. Epidemiological studies, not discussed here but reviewed elsewhere (*e.g.* Pope *et al.*,[1]), have demonstrated a clear relationship between the levels of PM$_{10}$ and exacerbations of asthma and COPD (chronic obstructive pulmonary disease), as well as deaths from cardiovascular causes, *i.e.* heart attacks and strokes.

2 Target Tissues for the Adverse Health Effects of PM$_{10}$

The range of mortality and morbidity described above indicates that there is a wide variety of tissues that are affected by PM$_{10}$ in ways that lead to disease and these are described in this section.

Airways

Both asthma and COPD are inflammatory diseases of the airways. The defences of the pulmonary airways comprise the mucociliary escalator, where mucus-secreting cells release mucus which traps deposited particles. Mucus with its trapped particles is then propelled upwards by ciliated cells to be either spat out or swallowed. In addition, the epithelial cells themselves are capable of responding with the release of inflammatory mediators to particle stimulation (*e.g.* Driscoll *et al.*[2]). Macrophages are also present in the airway walls and on the surfaces of the airways and these can phagocytose particles and release mediators. Within the airway walls are smooth muscle cells and mesenchymal cells which could also be targets for particles.

---

[1] C. A. Pope, D. V. Bates and M. E. Raizenne, *Environ. Health Perspect.*, 1995, **103**, 472.
[2] K. E. Driscoll, J. M. Carter, D. G. Hassenbein and B. Howard, *Environ. Health Perspect.*, 1997, **105**, 1159.

Terminal Airways and Proximal Alveoli

Particles deposit in large numbers beyond the ciliated airways in the terminal airways and proximal alveoli[3] where the net flow of air is zero and where, for very small particles, deposition efficiency increases because of the high efficiency of deposition by diffusion.[4] In this region it is the macrophages that play the most important role in removing particles. Macrophages phagocytose particles and eventually they migrate to the start of the mucociliary escalator and leave the lung with their cargo of particles, bound for the gut. Although some adverse effects associated with PM_{10} are clearly focused on the larger airways, effects beyond the ciliated airways could be important in the cardiovascular effects and in holding up neutrophils in the pulmonary vasculature which may be significant in causing lung injury (see later).

The Pulmonary Interstitium and Lymph Nodes

If particles cross the epithelium and enter the lung interstitium they are no longer likely to be cleared by the normal processes and will either remain in the subepithelial regions close to key responsive cell populations, such as interstitial macrophages, fibroblasts and endothelial cells, or be taken to the draining lymph nodes. Interstitial inflammation is likely to be more potentially harmful than inflammation in the alveolar spaces. The effects of dust in the lymph node are not known but adjuvant effects might be anticipated.

The Liver

Cardiovascular deaths are an important aspect of the adverse health effects of PM_{10}.[1] Classically, these are caused by the production of clots in the coronary vessels in the case of heart attacks and in the brain microvasculature in the case of stroke. Whilst intuitively an effect of inhaled particles on the lungs might be understandable, a link between the deposition of particles in the airways and effects that increase the likelihood of clots is more problematic. However, we have hypothesized that the inflammation arising in the lungs of persons inhaling PM_{10} could impact on the coagulation systems *via* the local production of pro-coagulant factors in the lungs or effects of mediators from the lungs on the liver, acting to increase the level of pro-coagulant factors. This is demonstrated diagrammatically in Figure 1. Research is underway to test this hypothesis experimentally, but there is epidemiological evidence to show that there is increased blood viscosity in a population during high pollution episodes;[5] fibrinogen is a major contributor to plasma viscosity. In this study there was an increase in the number of individuals in the highest quintile of blood viscosity, during a period of high air pollution.

[3] A. R. Brody, D. B. Warheit, L. Y. Chang, M. W. Roe, G. George and L. H. Hill, *Ann. N. Y. Acad. Sci.*, 1984, **428**, 108.

[4] P. J. Anderson, D. J. Wilson and A. Hirsch, *Chest*, 1990, **97**, 1115.

[5] A. Peters, A. Doring, H. E. Wichmann and W. Koenig, *Lancet*, 1997, **349**, 1582.

Figure 1 Diagram of the hypothetical scheme of events leading from deposition of PM$_{10}$ to morbidity and mortality outcomes (shaded ellipses)

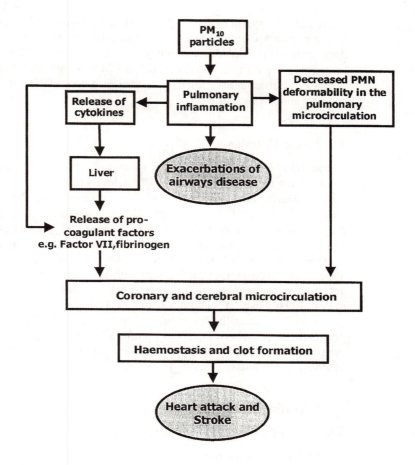

PMN in the Pulmonary Microvasculature

PMN (polymorphonuclear neutrophil leukocytes) are important mediators of lung injury in chronic inflammation, as a result of their ability to release injurious substances such as proteases and reactive oxygen species. These toxic products of the PMN could be released whilst the PMN are in the vascular space as well as when they extravasate into tissue. Neutrophils are known to be held up (or sequester) in the pulmonary microcirculation under normal circumstances owing to the fact that they have to deform because of their larger size in order to negotiate the smaller pulmonary capillary segments.[6] PMN deformability is therefore a critical initiating factor, prior to increased PMN-endothelial adhesion in PMN sequestration in the pulmonary microvasculture.[7] Any factors which change cell deformability will change PMN sequestration in the pulmonary circulation.[7] Airway inflammation, such as occurs in exacerbations of COPD, causes decreased PMN deformability, and hence increased PMN sequestration

[6] C. Selby and W. MacNee, *Exp. Lung Res.*, 1993, **19**, 407.
[7] C. Selby, E. Drost, P. K. Wraith and W. MacNee, *J. Appl. Physiol.*, 1991, **71**, 1996.

in the pulmonary microvasculature,[8] associated with evidence of oxidant stress.[9] These events are mediated by oxidative stress from acute smoking,[10] causing decreased neutrophil deformability.[11,12] As discussed below, oxidative stress from PM_{10} could also cause decreased PMN deformability, which could have important consequences for haemostasis and coagulation in the coronary and cerebral microvasculature.

3 Toxic Potency of PM_{10}

The levels of PM_{10} associated with adverse effects are very low and indeed some studies have shown no increase in PM_{10} levels that does not have an associated adverse health effect (*e.g.* Pope *et al.*,[1]), *i.e.* no threshold. The relative toxicity of PM_{10} is illuminated by considering the regulation of low toxicity or 'nuisance' particles in workplaces in the UK. The average PM_{10} level in UK urban areas is below 50 μg m^{-3} and levels are even less in country areas. By contrast, nuisance dusts are regulated in UK workplaces to 4 mg m^{-3} of respirable dust. Since PM_{10} appears to have adverse health effects without a threshold in some studies even at very low levels as measured by mass, this suggests that PM_{10} is a highly toxic material. However, the components of PM_{10} taken on their own are not particularly toxic at the levels present in ambient air, containing typically large proportions of carbon, salts and metals, as well as organic components.[13] The potential role of transition metals and very small particles in mediating the effects of PM_{10} are discussed below at length.

4 Ultrafine Particles

Toxicity of Ultrafine Particles

Research on ultrafine particles, *i.e.* those in the size range < 100 nm in diameter, provides a possible explanation of the toxicity of PM_{10}. Ultrafine particles are highly toxic to the lung, even when the particles are formed from materials that are not toxic as larger but still respirable particles, *e.g.* titanium dioxide and carbon. For example, the study of Ferin *et al.*[14] reported the bronchoalveolar lavage inflammatory profile in rats exposed to the same airborne mass concentration of TiO_2 as fine particles, approximately 250 nm in diameter (fine particles are hereafter defined as respirable particles that are larger than ultrafine particles), or ultrafine particle, 20 nm in diameter. Although rats were exposed to aerosols of 23 mg m^{-3} of both ultrafine TiO_2 and fine TiO_2, there was a marked

8   C. Selby, E. Drost, E. Lannan, P. K. Wraith and W. MacNee, *Am. Rev. Respir. Dis.*, 1991, **143**, 1359.
9   I. Rahman, D. Morrison, K. Donaldson and W. MacNee, *Am. J. Respir. Crit. Care Med.*, 1996, **154**, 1055.
10   W. MacNee, B. Wiggs, A. S. Belzberg and J. C. Hogg, *New Eng. J. Med.*, 1989, **321**, 924.
11   E. M. Drost, C. Selby, S. Lannan, G. D. O. Lowe and W. MacNee, *Am. J. Respir. Cell Mol. Biol.*, 1992, **6**, 287.
12   E. Dorst, C. Selby, M. M. E. Bridgeman and W. MacNee, *Am. Rev. Respir. Dis.*, 1993, **148**, 1277.
13   P. S. Gilmour, D. M. Brown, T. G. Lindsay, P. H. Beswick, W. MacNee and K. Donaldson, *Occup. Environ. Med.*, 1996, **53**, 817.
14   J. Ferin, G. Oberdorster and D. Penney, *Am. J. Respir. Cell Mol. Biol.*, 1992, **6**, 535.

inflammatory response seen with the ultrafine TiO_2 but little effect of the fine TiO_2. Thus the same material as ultrafine particles and as fine particles showed a dramatic difference in pathogenicity. We have reported a similar finding with fine (260 nm diameter) and ultrafine (14 nm diameter) carbon particles.[15] This suggests that ultrafine particles have a toxicity that is the result of their small size, as opposed to their chemical composition. The potential mechanisms of the toxicity of ultrafine particles has recently been reviewed.[16]

Deposition of Ultrafine Particles

The deposition fraction is high for ultrafine particles, approaching 50% for 20 nm particles, and interestingly the deposition efficiency is greater in a susceptible population (COPDs) than in normals.[4] This could be explained by the slower breaths of COPD patients that allow a longer residence time for the particles which would favour deposition that depends largely on Brownian motion, as is the case for these very small particles.[4]

Evidence that PM$_{10}$ Contains Ultrafine Particles

There is ample evidence that PM_{10} contains an ultrafine component,[17] and indeed, there is one report that decrements in the evening peak flow in a group of asthmatics were best associated with the ultrafine component of the airborne particles during pollution episodes.[18] Diesel exhaust is composed of singlet particles of around 50 nm diameter[19] and several studies have described a substantial component, in number terms of particles in the ultrafine range, although these represent a relatively small fraction of the total mass.[18]

Classical Particle Overload

The rat has been shown to be a high responder to particles, and indeed, dusts considered to be 'non-toxic' in humans (*e.g.* carbon, diesel particles, TiO_2) can cause severe lung injury, culminating in fibrosis, epithelial hyperplasia, metaplasia and cancer in rats, if exposure is to high airborne concentrations such that a high lung dose is attained.[20] This phenomenon of *overload* lung injury appears to be confined to rats and is not seen in other rodents even at similar lung doses (see papers in Mauderly and McCunney).[21] The phenomenon of overload is associated with slowed clearance from the deep lung (macrophage-mediated clearance) and subsequent rapid accumulation of dose with ongoing exposure, culminating in the effects described above. The original hypothesis for the mechanism of overload was focused on the volume of particles within the lung

[15] X. Y. Li, P. S. Gilmour, K. Donaldson and W. MacNee, *Thorax*, 1996, **51**, 1216.

[16] K. Donaldson, X. Y. Li and W. MacNee, *Aerosol Sci.*, 1998, in press.

[17] G. Oberdorster, R. Gelein, J. Ferin and B. Weiss, *Inhalation Toxicol.*, 1995, **71**, 111.

[18] A. Peters, H. E. Wichmann, T. Tuch, J. Heinrich and J. Heyder, *Respir. Crit. Care Med.*, 1997, **155**, 1376.

[19] R. L. Maynard and R. E. Waller, *Thorax*, 1996, **51**, 1174.

[20] J. L. Mauderly, *Inhalation Toxicol.*, 1996, **8** (suppl.), 1.

[21] J. L. Mauderly and R. J. McCunney. Proceedings of a conference held at the Massachusetts Institute of Technology, March 1995, *Inhalation Toxicol.*, 1996, **8** (suppl.).

and specifically the load volume of particles inside macrophages.[22] This hypothesis suggested that when the macrophages had phagocytosed a volume of particle equivalent to 6% of their internal volume, they began to show impaired ability to move and carry their particle burden to the start of the mucociliary escalator for removal from the lungs. Morrow[22] also calculated that by the time the average volume of particles inside macrophages reaches 60% of the total macrophage volume, their ability to move, and hence clearance, is completely inhibited; this has been confirmed.[23] Recently, however, new data have allowed a revision of this hypothesis, concluding that overload is best correlated to the metric of *surface area* of particles in the rat lung, not mass, volume or number of particles.[24,25] A role for surface area appears intuitively likely for *toxic* particles since the interaction between particles and biological systems will occur with the surface, not the internal mass, of the particle. However, it is not immediately apparent why *non-toxic* particles might mediate their effects *via* their surface. However, leaching of soluble components, including metals, will be greater from a large surface area and the potential role that this might play is described below.

Ultrafine Particles and Overload

As the particle diameter reduces for a constant mass of monodispersed (single diameter) particles, then the surface area increases dramatically.[17] It would appear, for this reason, that ultrafine particles may be more likely to cause overload at any given mass burden in the lungs because of their large surface area per unit mass. Amongst the most active of the 'low toxicity' dusts in causing lung overload tumours are the ultrafine particles,[25] presumably because they represent the biggest surface area per unit mass. Macrophages appear to be more adversely affected by loading of ultrafine TiO_2 than fine TiO_2. This is shown by the fact that the retention time of a radioactive marker particle in the lungs of rats exposed to ultrafine TiO_2 following inhalation exposure was inceased about eightfold compared to controls; however, lung burden data indicated that the macrophages were loaded with particles to a calculated volume of only 2.6%.[23] By contrast, the calculated macrophages load volume was 9% for fine TiO_2 and this caused only a doubling of the retention time of the test particle, in keeping with Morrow's prediction on the volumetric index of dose. This supports the contention that volume is not the dose metric that best predicts impairment of macrophage function caused by ultrafine particles; particle number or particle surface area may be the most important index for ultrafine particles. Macrophage functions associated with clearance are substantially impaired when the cells contain small load volumes of ultrafine particles, although this could be a large surface area.

The term overload should not be applied to the effect of ultrafine TiO_2 in inhibiting clearance, since it occurs at low lung burden. In relation to human risk assessment it should be noted that the phenomenon of overload occurs in rats

[22] P. E. Morrow, *Fundam. Appl. Toxicol.*, 1988, **10**, 369.
[23] G. Oberdorster, J. Ferin and B. E. Lehnert, *Environ. Health Perspect.*, 1994, **102** (suppl. 5), 173.
[24] G. Oberdorster, *Inhalation Toxicol.*, 1996, **8** (suppl.), 73.
[25] K. E. Driscoll, *Inhalation Toxicol.*, 1996, **8** (suppl.), 139.

and it is not clear whether it occurs in humans. In rats it occurs only at very high airborne mass concentrations, whereas PM$_{10}$ toxicity occurs in humans at remarkably low airborne mass concentrations. Classical overload is not, therefore, the mechanism of lung injury caused by PM$_{10}$. Whereas the most important factors contributing to slowed clearance in classical overload is the high macrophage burdens of particles, this is clearly not the case with ultrafines and toxicity to the macrophages may be more important. Ultrafine particles can be highly toxic to the lungs, as shown by high levels of LDH in the lungs of exposed rats,[15,26] and toxic particles may cause slowed clearance by a mechanism involving frank toxicity to macrophages.

Particle Numbers

Macrophages attempting to phagocytose a large number of ultrafine particles could be stimulated, by the high particle load, to release inflammatory mediators. In addition, the large numbers of particles may exceed the ability of the macrophages to phagocytose them, resulting in sustained stimulation of epithelial cells. This could cause the release of chemokines such as IL-8/MIP-1α that would contribute to inflammation. Increased production of these chemokines has been demonstrated in rats inhaling ultrafine carbon black[27] and also ultrafine particles of perfluoropolymer.[28]

Transfer of Particles to the Interstitium

Anything that interferes with the normal process of phagocytosis and macrophage migration to the mucociliary escalator can lead to the adverse outcome of *interstitialization*.[22] Interstitialization is an adverse outcome because interstitial particles cannot now be cleared *via* the normal pathways and must either remain in the interstitium, where they can chronically stimulate interstitial cells, or transfer to the lymph nodes. Interstitialization of particles was a prominent correlate of the onset of inflammation for ultrafine TiO$_2$ in the study of Ferin *et al.*,[14] and interstitialization of particles was found to arise concomitantly with overload inflammation.[29] Interstitialization is likely to occur when there is failed clearance, which could result from (a) particle-mediated macrophage toxicity or impairment of macrophage motility or (b) overload. Both of these events would allow increased interaction between particles and the epithelium that would favour interstitialization, and this could be further enhanced by increased epithelial permeability. In studies with rats it is clear that ultrafine particles and PM$_{10}$ can cause increased epithelial permeability.[15]

[26] Q. Zhang, Y. Kusake, K. Sato, A. Morita, K. Nakakuki, B. Li, K. Okada and K. Donaldson, *Toxicol. Appl. Pharmacol.*, 1998, in press.

[27] K. E. Driscoll, J. M. Carter, B. W. Howard, D. G. Hassenbein, W. Pepelko, R. B. Baggs and G. Oberdorster, *Toxicol. Appl. Pharmacol.*, 1996, **136**, 372.

[28] C. J. Johnston, J. N. Finkelstein, R. Gelein, R. Baggs and G. Oberdorster, *Toxicol. Appl. Pharmacol.*, 1996, **140**, 154.

[29] J. H. Vincent and K. Donaldson, *B. J. Ind. Med.*, 1990, **47**, 302.

5 Transition Metals

Transition Metals and Free Radicals in Particle Toxicity

Production of free radicals in the lung has been seen as a general mechanism mediating the biological activity of a number of different pathogenic particles[30,31] such as quartz,[32,33] coalmine dust,[34] residual oil fly ash,[35] asbestos[36] and synthetic mineral fibres.[37] The oxidative stress is considered to arise first from the particles themselves, normally by localized release of high concentrations of transition metals supplemented by the inflammatory cell influx that results from the primary interaction between lung cells and particles. Oxidative stress is a general signalling mechanism within cells that produces the transcription of a number of pro-inflammatory genes for cytokines, anti-oxidant enzymes, receptors and adhesion molecules.[38] Under the influence of oxidative stress, NF-κB separates from its inhibitor IκB and translocates to the nucleus to bind to the promoter region of key genes, allowing their transcription.[39] The ultrafine component of PM_{10}, with its large surface area, could generate free radicals that would be a substantial stimulus to transcription.

Free Radical Production by PM_{10}

To test this free radical hypothesis in the case of PM_{10} we have collected PM_{10} in Edinburgh and London. PM_{10} was found to have the ability to generate hydroxyl radical activity as shown in a supercoiled plasmid DNA scission assay[13] and by the ability to form the hydroxylated derivative of salicyclic acid (2,3-dihydroxybenzoic acid).[40] PM_{10} contained a large proportion of iron and the generation of hydroxyl radicals was blocked with iron chelators, confirming that Fenton chemistry is indeed the source of the hydroxyl radicals.[13] The majority of the available iron was in the form of Fe^{3+}, but the presence in the lung of reductants such as superoxide anion and glutathione (GSH) would be able to initiate the reaction by reducing Fe^{3+} to Fe^{2+}. Following the instillation of PM_{10} into the lungs of rats there was evidence of inflammatory neutrophil influx and oxidative stress, as shown by depletion of GSH in lung lining fluid.[15] Importantly, PM_{10} caused significantly more inflammation than a similar mass

[30] T. P. Kennedy, R. Dodson, N. V. Rao, H. Ky, C. Hopkins, M. Baser, E. Tolley and J. R. Hoidal, *Arch. Biochem. Biophys.*, 1989, **269**, 359.

[31] K. Donaldson, P. H. Beswick and P. S. Gilmour, *Toxicol. Lett.*, 1996, **88**, 293.

[32] V. Castranova, V. Vallyathan, D. M. Ramsey, J. L. McLaurin, D. Pack, S. Leonard, M. W. Barger, J. C. Ma, N. S. Dalal and A. Teass, *Environ. Health Perspect.*, 1997, **105**, 1319.

[33] K. Zay, D. Devine and A. Churg, *J. Appl. Physiol.*, 1995, **78**, 53.

[34] N. S. Dalal, M. M. Suryan, V. Vallyathan, F. Y. Green, B. Jafari and R. Wheeler, *Ann. Occup. Hyg.*, 1989, **33**, 79.

[35] K. L. Dreher, R. H. Jaskot, J. R. Lehmann, J. H. Richards, J. K. Mcgee, A. J. Ghio and D. L. Costa, *J. Toxicol. Environ. Health*, 1997, **50**, 285.

[36] P. S. Gilmour, P. H. Beswick, D. M. Brown and K. Donaldson, *Carcinogenesis*, 1995, **16**, 2973.

[37] D. M. Brown, C. Fisher and K. Donaldson, *J. Toxicol. Environ. Health Part A*, 1998, **53**, 101.

[38] I. Rahman and W. MacNee, *Thorax*, 1998, in press.

[39] M. Meyer, H. K. Pahl and P. A. Bauerle, *Chem. Biol. Interact.*, 1994, **91**, 91.

[40] K. Donaldson, D. M. Brown, C. Mitchell, M. Dineva, P. H. Beswick, P. Gilmour and W. MacNee, *Environ. Health Perspect.*, 1997, **105** (suppl. 5), 1285.

(125 μg) of carbon black that was 260 nm in diameter, *i.e.* not in the ultrafine size range; the inclusion of a mass bolus control is vital to interpreting this type of data.

ROFA (residual oil fly ash) has been used as a surrogate for PM$_{10}$, although it is very different in many respects from PM$_{10}$. ROFA has been found to cause pulmonary inflammation after instillation, *via* a transition metal-mediated mechanism.[35] Furthermore, in rats instilled with ROFA, an intraperitoneal injection of the free radical scavenger DMTU lowered the amount of PMN influx to the lung.[41] ROFA particles cause increased transcription of cytokine genes by human bronchial epithelial cells *in vitro via* a transition metal-mediated mechanism,[42] as shown by the fact that the effect could be blocked with the metal chelator deferoxamine. Interestingly, the stimulation of cytokine production could be mimicked by vanadium salts in solution but not by iron or nickel sulfate, pointing towards the possible importance of vanadium.

6 Hypothetical Mechanism for the Cardiovascular Effects of PM$_{10}$

The effect of PM$_{10}$ in increasing the numbers of cardiovascular deaths (*i.e.* from strokes and heart attacks) is one of the most puzzling of its adverse effects. We have suggested that the local pulmonary inflammation could be translated into increased pro-coagulant status and conditions that could favour haemostasis in the coronary and cerebral microvasculature; these conditions would promote heart attack and stroke, respectively. Two mechanisms could be operative: (a) increased production of pro-coagulant by the liver or the lungs; (b) decreased deformability of PMNs.

Lung and Liver as Sources of Pro-coagulant Factors

We have shown that following deposition of ultrafine particles in the lungs, there is inflammation, oxidative stress[15,43] and up-regulation of oxidative stress-responsive genes such as TNF, SOD and iNOS.[15,44] In preliminary studies we have shown that inhalation of ultrafine particles in the rat increases the levels of factor VII in plasma (unpublished data). Factor VII has been shown to be a risk factor for cardiovascular disease in population studies.[45,46] This association is supported by two recent studies. The first showed that polymorphisms in the factor VII gene influence both the plasma concentration of factor VII antigen and activity, and the risk of myocardial infarction;[47] secondly, the use of warfarin to lower factor VII plasma concentrations, to those associated with low vascular risk in epidemiological studies, is accompanied by significant protection from fatal vascular events.[48] The principal site of synthesis of factor VII is the

[41] J. A. Dye, K. B. Adler, J. H. Richards and K. Dreher, *Am. J. Respir. Cell Mol. Biol.*, 1997, **12**, 625.
[42] J. D. Carter, A. J. Ghio, J. M. Samet and R. B. Devlin, *Toxicol. Appl. Pharmacol.*, 1997, **146**, 180.
[43] X. Y. Li, W. MacNee and K. Donaldson, unpublished results.
[44] X. Y. Li, P. S. Gilmour, K. Donaldson, W. MacNee and A. Churg, *Am. J. Respir. Crit. Care Med.*, 1998, in press.
[45] T. W. Meade, V. Ruddock, R. Chakrabarti and G. J. Miller, *Lancet*, 1993, **342**, 1076.
[46] T. W. Meade, S. Mellows, M. Brozovic *et al.*, *Lancet*, 1986, **2**, 533.
[47] L. Iacovelli, A. Di Castelnuovo, P. de Knijff *et al.*, *New Engl. J. Med.*, 1998, **338**, 79.
[48] The Medical Research Council's General Practice Research Framework, *Lancet*, 1998, **351**, 233.

hepatocyte, but mRNA for both factor VII and tissue factor, which is also involved in coagulation, have been demonstrated in alveolar macrophages,[49] and these cells are capable of synthesis of factor VII *in vitro*.[50] Thus local inflammation in the lungs, and the activation of alveolar macrophages, could result in local and generalized release of pro-coagulant factors, which may enter the blood stream and have a systemic effect.

Local inflammation in the lungs could also result in increased pro-coagulant activity from the liver. There are at least two possible mechanisms by which these effects can occur following inhalation of particulate air pollution: (1) through the release of cytokines from inflammatory cells in the lungs; (2) as a result of the development of systemic oxidant stress. Both of these could up-regulate the genes for pro-coagulant factors in hepatocytes. We have shown clear evidence of systemic oxidant stress, measured as a decrease in the anti-oxidant capacity of the plasma following installation of PM_{10} particles, which was not apparent following instillation of the same mass of fine carbon particles.[43] Candidate cytokines which may result in up-regulation of pro-coagulant factors in the liver are $TNF\alpha$ and IL-6.

Alterations in Blood Rheology as a Cause of Increased Haemostasis

One unique feature of the pulmonary microcirculation is the close proximity of the distal airspace to the circulating blood, across the alveolar–capillary membrane, allowing easy access for inflammatory mediators in the airspaces to reach the blood. A recent study has shown increased plasma viscosity during an episode of air pollution, confirming that a 'signal' from the lungs following exposure to particles can affect plasma indices.[5] We have shown that the ability of neutrophils to deform in transit in the pulmonary capillaries is the factor which initiates neutrophil sequestration in the lungs, as the precursor of migration into the airspaces.[7] We have also shown that lung inflammation, which occurs during smoking or acute exacerbations of chronic obstructive pulmonary disease, increases neutrophil sequestration in the lungs,[8,10] very likely *via* an oxidant-induced decrease in cell deformability.[11] Thus oxidant-generating ultrafine environmental particles depositing in the distal airspaces may also produce increased neutrophil sequestration in the lungs, and so contribute to the initiation of lung inflammation. In addition, the decrease in neutrophil deformability induced in cells in transit in the pulmonary microcirculation, or the release of less deformable cells from the bone marrow in response to particle-induced lung inflammation,[51] may result in sequestration of these cells in the microcirculation of other organs, such as the heart and the brain, so contributing to local haemostasis and thrombotic events.

7 Activation of NF-κB in the Lungs after Inhalation of Ultrafine Particles as a Central Initiating Event

The transcriptional activator NF-κB is a nuclear factor of the Rel family that is

[49] M. P. McGee, R. Devlin, G. Saluta and H. Koren, *Blood*, 1990, **75**, 122.
[50] H. A. Chapman, C. L. Allen and O. L. Stone, *J. Clin. Invest.*, 1985, **75**, 2030.
[51] T. Terashima, B. Wiggs, D. English, J. C. Hogg and S. F. van Eeden, *Am. J. Respir. Crit. Care Med.*, 1997, **155**, 1441.

translocated to the nucleus to permit expression of a wide range of pro-inflammatory genes.[39] The NF-κB heterodimer, comprising p65 and p50 proteins, is found in resting cells bound to its inhibitor IκB, which masks the nuclear translocation signal and so prevents its translocation to the nucleus. Under oxidative stress or a range of other stimuli such as TNF, the IκB is phosphorylated and then degraded *via* the ubiquitin proteosome system, allowing the NF-κB to relocate to the nucleus.[52] Genes that have a κB binding site in their promoter include cytokines, growth factors, chemokines, and adhesion molecules and receptors.[38] In addition, the genes for tissue factor contain an NF-κB binding site[53] and so may be susceptible to transcriptional activation during oxidative stress. We have demonstrated activation of NF-κB by oxidative stress in airspace epithelial cells.[54] The deposition of particles that deliver an oxidative stress to the lungs cause activation of NF-κB, and possibly other oxidative stress-responsive transcription factors, that initiate a cascade of gene expression which lead to a pro-coagulant and haemostatic state.

8 Implications of an Oxidative Stress-mediated Mechanism of Action of PM$_{10}$ for Susceptibility in Patients with Airways Disease

The Central Role of the Epithelium

Because of the deposition of particles on the epithelium, prior to phagocytosis, it seems likely that the epithelium is a target for the PM$_{10}$ in leading to increased asthma and COPD attacks. Antigens for asthma are present in most atmospheres and to trigger an asthma attack the antigen need only gain access to the subepithelial lymphoid tissue. There is evidence that various kinds of environmental particles such as ROFA,[41] PM$_{10}$[15] and also ultrafine carbon black[55] can compromise the epithelium by causing injury or oxidative stress. This presents the possibility that increased production of inflammatory mediators and increased permeability to antigens may be a mechanism for the induction of asthma attacks, additional to the fact that the underlying inflammation in the airways of asthmatics means that they are in a 'primed' state for the further oxidative stress caused by depositing PM$_{10}$.

Existing Oxidative Stress in Susceptible Populations

It should be noted that the principal pulmonary effects of PM$_{10}$ are seen in susceptible populations, such as those with airways disease. If, as hypothesized here, the PM$_{10}$ has its effect mainly by a mechanism that involves oxidative stress, then these susceptible populations might be susceptible because of pre-existing oxidative stress. We have utilized an assay (Trolox Equivalent Antioxidant Defence; TEAC) that detects the global anti-oxidant defence in the plasma, and

[52] K. Brown, S. Gerstberger, L. Carlson, G. Franzoso and U. Siebenlist, *Science*, 1995, **267**, 1485.
[53] C. L. Orthner, G. M. Rodgers and L. A. Fitzgerald, *Blood*, 1995, **86**, 436.
[54] B. Mulier, T. Watchorn and W. MacNee, *Am. J. Respir. Crit. Care Med.*, 1998, in press.
[55] V. Stone, W. MacNee, S. Faux and K. Donaldson, *Toxicol. in vitro*, 1998, submitted for publication.

have demonstrated depleted anti-oxidant defences in patients with airways disease.[9] Plasma samples from patients with asthma and COPD were found to have significantly lower TEAC values than normal patients and these were found to be further lowered during asthma attacks and exacerbations of COPD. Clearly these patients would be susceptible to an oxidative insult such as that hypothesized here to emanate from PM_{10}.

Chemical Carcinogens

JOHN C. LARSEN AND POUL B. LARSEN

1 Introduction

Although the dominant risk factor in lung cancer is cigarette smoking, epidemiological studies over the last half century have suggested that general ambient air pollution may contribute to increased rates of lung cancer. This evidence is derived from comparisons of urban and rural populations in a number of cohort and case-control studies. Studies on occupational groups have provided strong supportive evidence.[1-3] Cohort studies from several countries have shown relative risks of lung cancer in urban areas of the order of 1.5 or lower, when adjusted for smoking. These findings mainly relate to male smokers. In a number of case-control studies, the evidence of increased relative risks for lung cancer in urban areas was also recorded primarily in male smokers. Increased risks for females and non-smokers in urban areas have only been indicated in a few studies, because the number of cases were too small to provide meaningful differences.[3] A smoking-standardized relative lung cancer risk of 1.5 was found among men and of 1.2 among women living in highly polluted areas of Poland.[4] It was estimated that 4.3% of the lung cancers in men and 10.5% in women could be attributed to air pollution.

Ambient air in densely populated urban areas contains a variety of organic chemicals, including known carcinogens such as benzo[a]pyrene (BaP) and benzene. Humans are thus exposed to complex mixtures which also include carbon-based particles that absorb organic compounds, oxidants such as ozone, sulfuric acid in aerosol form, and inorganics such as arsenic and chromium. In this paper, only the organic chemicals which most often have been implicated in cancer risk from general air pollution are considered, notably benzene, 1,3-butadiene, formaldehyde and, in particular, polynuclear aromatic hydrocarbons (PAHs).

Chemical carcinogenesis is a multistage process including initiation, promotion

[1] A. J. Cohen and C. A. Pope III, *Environ. Health Perspect.*, 1995, **103**, 219.
[2] G. Pershagen and L. Simonte, in *Air Pollution and Human Cancer*, ed. L. Tomatis, Springer, Heidelberg, 1990, p. 63.
[3] K. Hemminki and G. Pershagen, *Environ. Health Perspect.*, 1994, **102**, 187.
[4] W. Jedrychowski, H. Becher, J. Wahrendorf and Z. Basa-Cierpialek, *J. Epidemiol. Commun. Health*, 1990, **44**, 114.

and progression. In the initiation step, which may involve procarcinogen activation to genotoxic species, a genotoxic agent interacts with DNA and induces mutations in the target cells. If not repaired by the endogenous DNA repair systems the mutations may be fixed in initiated daughter cells. During promotion the initiated cells experience growth advantages, resulting in increased cell proliferation and the development of (benign) neoplasia. The promotion can be stimulated by many chemicals acting on enzymes controlling cell homeostasis, by respiratory tract irritants and by tissue injury and inflammation. Promotion is not thought to involve genotoxic mechanisms. During the progression phase the cells become malignant and capable of invading other tissues (metastasis). Genotoxic events are thought to play a role during progression.

Data from epidemiological studies of occupationally exposed workers or animal studies may be used to estimate risks from chemicals at the much lower levels normally encountered in ambient air. From a theoretical point of view, there is no threshold dose for the carcinogenic effect of a genotoxic agent below which there is no risk. Risks at low levels of exposure have to be estimated by extrapolation. In most cases, linear dose–response relationships are assumed, and a unit risk factor can be calculated. Normally, unit risk is the probability (95% upper confidence limit) of developing cancer from continuous lifetime inhalation of $1~\mu g\,m^{-3}$ of the airborne chemical. In the case of compounds that produce cancer by non-genotoxic mechanisms, it is believed that threshold exposures for the carcinogenic effect can be established.

There seems to be a genetic predisposition in the individual susceptibility to lung cancer, at least for cigarette smoking. Of the many steps involved in carcinogenesis, most toxicological information on differences in susceptibility points to the proximate procarcinogen metabolism steps. The major classes of organic chemical carcinogens present in ambient air undergo oxidation (phase I) and conjugation steps (phase II). During phase I, intermediates can be bioactivated to potential carcinogens, while they are detoxified during phase II. Genetic polymorphisms seem to exist for both phase I and phase II enzymes, and an individual's susceptibility to lung cancer may be conferred by the balance between the capacity to activate inhaled procarcinogens to ultimate carcinogens and detoxify proximate/ultimate carcinogens.[5]

2 Benzene

Sources and Levels in Ambient Air

Benzene is a colourless liquid with a boiling point of $80\,^{\circ}C$ and a high vapour pressure. Benzene constitutes about 1–5% of gasoline. The major release of benzene into the environment (> 80%) comes from automobile exhaust, where benzene accounts for about 5% of the total hydrocarbon emissions.[6,7] Background levels of benzene at remote and rural areas have been reported at 0.5–$1.5~\mu g\,m^{-3}$.

[5] S. D. Spivack, M. J. Fasco, V. E. Walker and L. S. Kaminsky, *Crit. Rev. Toxicol.*, 1997, **27**, 319.
[6] IPCS, *Benzene*, Environmental Health Criteria 150, International Programme on Chemical Safety, World Health Organization, Geneva, 1993.
[7] L. Wallace, *Environ. Health Perspect.*, 1996, **104**, 1129.

Average levels of 5–15 μg m^{-3} have been found in Western European and North American cities. However, at the kerbside of busy streets, levels of 30 μg m^{-3} or higher have been measured.[6,8] Measurements of benzene concentration at 20 sites in California showed a decreasing time trend from an average level of about 9 μg m^{-3} in 1986 to about 4 μg m^{-3} in 1994.[7] Levels were found to be about twice as high in the winter compared to the summer. In Sweden, winter levels (1993/1994) in 22 towns were reported at 2.6–7.4 μg m^{-3}.[8]

Human Exposure

About 40% of the average daily benzene exposure of non-smokers can be attributed to benzene in outdoor air, while indoor air may account for about 31% (levels of about 3–8 μg m^{-3}). Driving a vehicle may account for about 19%, as the average level inside cars is about 40 μg m^{-3}.[7] At petrol stations, one may be exposed to a benzene level of about 1 mg m^{-3} during refuelling.[8] Cigarette smoke is an important additional source of benzene, and smokers are exposed to about 55 μg of benzene per cigarette. Further, cigarette smoke significantly contributes to indoor benzene levels, as the levels in homes of smokers are generally 3.5–4.5 μg m^{-3} higher than in homes with no tobacco smoke.[7]

Overall, inhalation of benzene accounts for about 98–99% of the total daily benzene exposure, and the average intake is estimated to about 200 μg for non-smokers and 2 mg for smokers. Oral ingestion, *i.e.* through food and drinking water, does not significantly contribute to benzene intake (< 1%).[7]

Toxicokinetics

Inhaled benzene is readily absorbed from the lungs. In humans at exposure levels of 163–326 mg m^{-3}, about 50% of the inhaled benzene was reported to be absorbed. The highest levels of benzene are obtained in lipid-rich tissues, and benzene is known to cross the placental barrier.[6,8] Benzene metabolism is similar in humans and experimental animals. Benzene is metabolized mainly in the liver by the cytochrome P450 2E1 enzyme system and involves formation of the unstable intermediate benzene oxide. The major metabolites formed are phenol, catechol and hydroquinone. Benzoquinone and muconaldehyde are minor metabolites considered to contribute to the toxicity of benzene. In rodents, the formation of these metabolites appears to be saturable, resulting in proportionally higher fractions of these toxic metabolites at lower levels. Mice were found to be most active in converting benzene to benzoquinone and muconaldehyde.[6,8]

In humans, benzene is eliminated unmetabolized with the expiratory air, whereas metabolites are excreted in urine, primarily as the sulfate and glucuronide conjugates of phenol.[6]

Toxicological Effects

The primary acute reponses to benzene by humans involve narcotic symptoms

---

[8] WHO, *Benzene*, in *WHO Air Quality Guidelies for Europe*, 2nd edn., World Health Organization, Regional Office for Europe, Copenhagen, 1998.

from the central nervous system, such as dizziness, headache, drowsiness and nausea. Headache, lassitude and weakness have been reported at benzene levels between 160 and 480 mg m^{-3}. However, short-term exposure to benzene at levels of 1600–3200 mg m^{-3} (500–1000 ppm) are tolerated by humans without serious adverse effects.[6,9]

The main toxic manifestations from repeated occupational exposure to benzene are bone marrow depression with anaemia, leucopenia or thrombocytopenia. The effects follow a clear dose–response relationship down to an exposure level of about 32 mg m^{-3} (10 ppm). It has been suggested that it is the combined effects of several of the benzene metabolites that adversely affect the functioning of bone marrow cells and lead to pancytopenia and aplastic anaemia.[6]

Many studies have reported structural and numerical chromosomal aberrations in lymphocytes and bone marrow cells from workers exposed to benzene, and in some cases at exposure levels as low as 4–7 mg m^{-3}. Benzene has not induced gene mutations *in vitro*, but in animal studies benzene exposures have caused a variety of chromosomal aberrations which support the view that benzene is a clastogenic agent.[8] The mechanism of action is not fully understood. Metabolic data suggest that several reactive metabolites may form adducts with DNA and proteins. Binding to protein components of the spindle apparatus may inhibit the mitosis of the cell, while binding to DNA may be an initiating event in carcinogenesis.[6]

Carcinogenicity

Results from carcinogenicity testing of benzene in experimental animals have revealed the induction of various types of lymphomas/leukaemias. However, most neoplasms found were of epithelial origin, *i.e.* in the Zymbal gland, liver, mammary gland and oro-nasal cavity. A carcinogenic effect was still found at 320 mg m^{-3} in inhalation studies and at a dose of 25 mg kg^{-1} of body weight per day by gavage. Thus, in experimental animals, benzene was not only a specific leukaemogenic agent as in humans (see below), but rather a multipotent carcinogen.[6]

Epidemiological Data

Several epidemiological studies have consistently shown a positive relationship between benzene exposure and leukaemia. The most thorough study involves a group of workers in the rubber film industry (the Pliofilm cohort). A total of 1212 workers employed between 1940 and 1965 were followed up to 1987. The cohort was unique because of the high quality data with respect to exposure history and medical surveillance. In the latest up-date a total of 14 cases of leukaemia and four cases of multiple myelomas had occurred. The standard mortality rate (SMR) for all lymphatic and haematopoietic cancers (SMR = 221) and for leukaemia (SMR = 360) were significantly increased. A strong positive trend was found between leukaemia mortality and increased cumulative exposure.[6,8]

[9] D. J. Paustenbach, R. D. Bass and P. Price, *Environ. Health Perspect.*, 1993, **101**, 177.

Evaluation and Risk Assessment

Based on the epidemiological evidence, benzene was considered a human carcinogen (group 1) by the IARC, the International Agency for Research on Cancer.[10]

The WHO, in its update of the Air Quality Guidelines for Europe, used data from the updated Pliofilm cohort and models based on relative risk and cumulative exposure to calculate unit risks for benzene in the range of 4.4–7.5×10^{-6} per $\mu g\,m^{-3}$, with a geometric mean of 6×10^{-6} per $\mu g\,m^{-3}$. Based on this, the WHO concluded that the average concentration of airborne benzene associated with an excess lifetime cancer risk of 10^{-6} was $0.18\,\mu g\,m^{-3}$.[8] Thus, at an average urban air level of 5–$15\,\mu g\,m^{-3}$ this risk estimate predicts 30–90 extra cases of leukaemia among 1 million exposed for their lifetime.

3 1,3-Butadiene

Sources and Levels in Ambient Air

1,3-Butadiene is a colourless gas, only slightly soluble in water. The two conjugated double bonds in the molecule make it highly reactive.[11,12]

Traffic exhaust is the primary source of 1,3-butadiene in ambient air. 1,3-Butadiene is emitted at an average vehicle rate of 5.6–$6.1\,mg\,km^{-1}$ and comprises roughly 0.35% of the hydrocarbons in the exhaust (exhaust contains about 44–$72\,\mu g\,m^{-3}$ 1,3-butadiene). In general, average long-term exposure levels of 1,3-butadiene in urban air are around $1\,\mu g\,m^{-3}$. The mean level in 19 cities in the USA was reported to be $1.4\,\mu g\,m^{-3}$.[11,12]

1,3-Butadiene is rapidly transformed in the atmosphere by reaction with hydroxyl radicals, ozone and nitrogen trioxide radicals. These reactions generate formaldehyde, acrolein and organic nitrates. In the daytime during the summer the residence time of 1,3-butadiene in the atmosphere is estimated to be less than one hour, while in the winter on cloudy days it may exceed one day.[11]

Human Exposure

The general public is exposed to concentrations of 1,3-butadiene in the low $\mu g\,m^{-3}$ range through ambient and indoor air. Indoor cigarette smoke may significantly contribute to the exposure (0.2–$0.4\,mg$ per cigarette in sidestream smoke) and smoky indoor levels of 10–$20\,\mu g\,m^{-3}$ have been reported. An

[10] IARC, *Benzene*, in *Overal Evaluations of Carcinogenicity: An Updating of IARC Monographs Volumes 1 to 42*, IARC Monographs on the Evaluation of Carcinogenic Risks to Humans, suppl. 7, International Agency for Research on Cancer, Lyon, 1987, p. 120.

[11] US EPA, *Motor Vehicle-Related Air Toxics Study*, Technical Support Branch Emission Planning and Strategies Division, Office of Mobile Sources, Office of Air and Radiation, US Environmental Protection Agency, 1993.

[12] WHO, *1,3-Butadiene*, in *WHO Air Quality Guidelines for Europe*, 2nd edn., World Health Organization, Regional Office for Europe, Copenhagen, 1998.

additional source of exposure is inhalation of gasoline vapours at petrol stations and inside cars.[12,13]

Toxicokinetics

In animal studies the degree of absorption following inhalation of 1,3-butadiene depends on the species tested and the exposure level used. At a (low) exposure level of $1.77\,mg\,m^{-3}$, 20% of the inhaled dose was retained by mice, while only 6% was retained by rats. At a higher exposure level of $2210\,mg\,m^{-3}$ (1000 ppm), 4% was retained by mice and 2.5% by rats.[14] Studies in monkeys indicate a lower uptake in this species than in rats and mice.[15]

Metabolic conversion to the genotoxic metabolites 1,2-epoxy-3-butene and 1,2:3,4-diepoxybutane by cytochrome P450 enzymes is regarded as the crucial event for the carcinogenic potential of 1,3-butadiene. Large species differences have been found with respect to the formation and degradation of these active metabolites. Thus, at a similar exposure level, considerably higher levels of 1,2-epoxy-3-butene and 1,2:3,4-diepoxybutane were measured in the blood from mice than from rats.[16] This may explain why the mouse is more susceptible to the carcinogenic effect of 1,3-butadiene than the rat. There is evidence that primates generate 1,2-epoxy-3-butene to a lesser extent than rodents; however, the extent to which 1,2:3,4-diepoxybutane is produced in primates is not known.[16]

Toxicological Effects

1,3-Butadiene is of low acute toxicity in experimental animals with LD_{50} values above $221\,000\,mg\,m^{-3}$ (100 000 ppm). Exposure levels of several thousand ppm may cause respiratory tract and eye irritation in humans. In long-term experiments, 1,3-butadiene exposure has caused testicular atrophy in male mice at $1380\,mg\,m^{-3}$ and ovarian atrophy in female mice at $13.8\,mg\,m^{-3}$ and higher levels. In reproductive and developmental studies, decreased foetal weight (exposure level $88.4\,mg\,m^{-3}$), extra ribs ($442\,mg\,m^{-3}$) and sperm head abnormalities ($2210\,mg\,m^{-3}$) have been reported in mice. In the rat, skeletal variations ($2210\,mg\,m^{-3}$) and skeletal abnormalities ($17\,600\,mg\,m^{-3}$) have been observed.[13]

1,3-Butadiene and its metabolites, 1,2-epoxy-3-butene and 1,2:3,4-diepoxybutane, have shown genotoxicity in a variety of *in vitro* and *in vivo* studies. 1,3-Butadiene was particularly active in studies using mice, owing to the high capacity of this species to bioactivate 1,3-butadiene to the genotoxic metabolites.[13,17]

[13] M. W. Himmelstein, J. F. Acquavella, L. Recio, M. A. Medinsky and J. A. Bond, *Crit. Rev. Toxicol.*, 1997, **27**, 1.

[14] J. A. Bond, A. R. Dahl, R. F. Henderson, J. S. Dutcher, J. L. Mauderly and L. S. Birnbaum, *Toxicol. Appl. Pharmacol.*, 1986, **84**, 617.

[15] A. Dahl, J. D. Sun, L. S. Birnbaum, J. A. Bond, W. C. Griffith, J. L. Mauderly, B. A. Muggenburg, P. J. Sabourin and R. F. Henderson, *Toxicol. Appl. Pharmacol.*, 1991, **110**, 9.

[16] R. F. Henderson, J. R. Thornton-Manning, W. E. Bechtold and A. Dahl, *Toxicology*, 1996, **113**, 17.

[17] IARC, *1,3-Butadiene*, IARC Monographs on the Evaluation of Carcinogenic Risks to Humans, vol. 54, International Agency for Research on Cancer, Lyon, 1992, p. 237.

Carcinogenicity

1,3-Butadiene has been tested for carcinogenicity in one rat study and four mouse studies using inhalation exposures in the range of 14 to $17\,600\,mg\,m^{-3}$. In all studies and at all exposure levels, increased incidences of tumours were found.

In rats tested at 2200 and $17\,600\,mg\,m^{-3}$ (6 h per day, 5 days per week for up to 111 weeks), females had exposure-related increased rates of mammary gland adenomas/carcinomas, thyroid follicular cell adenomas, uterine sarcomas and Zymbal gland sarcomas, while males had increased incidences of testicular Leydig cell tumours and pancreatic exocrine adenomas.[13] The mouse was found more susceptible to the carcinogenic effects of 1,3-butadiene exposure than the rat. Exposure at levels of 13.8, 44.2, 138, 442 and $1380\,mg\,m^{-3}$ (6 h per day, 5 days per week for up to 2 years) resulted in treatment-related increased incidences of lymphocytic lymphomas, haemangiosarcomas of the heart, and neoplasms of the lung, forestomach, Harderian gland, preputial gland, liver, mammary gland and ovary. Significant increases in lung adenomas/carcinomas were observed down to the lowest exposure level of $13.8\,mg\,m^{-3}$.[18]

Epidemiological Data

The most important epidemiological data include cohort mortality studies of butadiene monomer (BDM) workers and styrene butadiene rubber (SBR) workers. In a recent update of a cohort of 2795 BDM workers with occupational exposure solely to 1,3-butadiene, an increased standardized mortality rate (SMR) of 147 (95% confidence limit: 106–198) was found for lymphohaematopoietic cancers (LHC). However, many of the LHC cases were seen in workers engaged fewer than five years at the plant and therefore might have been exposed to other agents in previous occupations. Furthermore, only very rough exposure indices were available as measurements of 1,3-butadiene occupational levels were not performed.[19]

An increased SMR of 131 (95% confidence limit: 97–174) for leukaemia was found in a recent update of a cohort of 15 649 SBR workers from seven plants. Even higher SMRs were found for leukaemia in subgroups with a long duration of employment, and for workers occupied at sites and processes with high levels of styrene and 1,3-butadiene.[20] From estimated individual cumulative exposure indices, a positive relationship was found between 1,3-butadiene exposure and increased SMR values for leukaemia.[21] In these studies, no increased SMR was found for other types of lymphohaematopoietic cancer.

[18] R. L. Melnick, J. Huff, B. J. Chou and R. A. Miller, *Cancer Res.*, 1990, **50**, 6592.

[19] B. J. Divine and C. M. Hartman, *Toxicology*, 1996, **113**, 169.

[20] E. Delzell, N. Sathiakumar, M. Hovinga, M. Macaluso, J. Julian, R. Larson, P. Cole and D. Muir, *Toxicology*, 1996, **113**, 182.

[21] M. Macaluso, R. Larson, E. Delzell, N. Sathiakumar, M. Hovinga, J. Julian, D. Muir and P. Cole, *Toxicology*, 1996, **113**, 190.

Evaluation and Risk Assessment

The IARC has evaluated 1,3-butadiene as probably carcinogenic to humans (group 2A), based on sufficient evidence for carcinogenicity from experimental animal data and limited evidence for carcinogenicity from human data.[17]

Owing to uncertainties as to which animal species and extrapolation models should be used for risk assessment, the WHO in its update of the Air Quality Guidelines for Europe did not recommend a risk estimate for 1,3-butadiene as a basis for a guidance value. The WHO expressed the view that despite the fact that there is some evidence of the carcinogenicity of 1,3-butadiene in humans, albeit equivocal, decisions regarding ambient air standards should be made with prudence.[12]

In a review of seven risk assessments of 1,3-butadiene performed by different agencies and organizations, the calculated risk estimates for an occupational level of 1 ppm (2.21 mg m^{-3}) 1,3-butadiene ranged from 0 to 2613 extra deaths per 10 000 exposed workers. This reflects the large differences between the assessments in selecting and utilizing experimental data (using rat or mouse data, and how to scale from animal dose to human dose) and in the use of models for low-dose level extrapolation. In general, considerably lower risk estimates were obtained when the data from the rat study were used. However, at present there is no general agreement upon which experimental animal species (rat or mouse) should be used for the extrapolation to humans.[13]

4 Formaldehyde

Sources and Ambient Air Levels

Formaldehyde is a very water soluble and reactive gas with a pungent odour. Formaldehyde is formed by photooxidation of hydrocarbons in the troposphere, where naturally occurring methane is the most important source. Background levels of formaldehyde are below 1 μg m^{-3} [22,23] Traffic emissions are by far the most important sources of formaldehyde in ambient air in urban areas. Gasoline engines may emit several hundred mg of formaldehyde per litre of combusted fuel and diesel engines about 1 g formaldehyde per litre of fuel. The use of exhaust catalytic converters reduces the formaldehyde emission to less than one-tenth of those levels.[22] Levels in urban areas with anthropogenic hydrocarbon and aldehyde emissions from traffic are reported to be 1–20 μg m^{-3}. However, peak levels up to 100 μg m^{-3} may occur during severe inversion episodes.[22,23]

In air, formaldehyde photolyses and reacts rapidly with free radicals. The half-life in sunlight is a few hours. Owing to the high water solubility, formaldehyde may also be eliminated with rain. Levels in rainwater have been reported to be 0.1–0.2 mg kg^{-1}.[22]

[22] IPCS, *Formaldehyde*, Environmental Health Criteria 89, International Programme on Chemical Safety, World Health Organization, Geneva, 1989.

[23] IARC, in *Wood Dust and Formaldehyde*, IARC Monographs on the Evaluation of the Carcinogenic Risk of Chemicals to Humans, vol. 62, International Agency for Research on Cancer, Lyon, 1995, p. 217.

Human Exposure

In humans and other mammals, formaldehyde is an essential metabolic intermediate in all cells, *e.g.* in the biosynthesis of purines, thymidine and certain amino acids. The concentration of formaldehyde in the blood of humans not exposed to exogenous formaldehyde has been reported to be $2.6\,mg\,kg^{-1}$.[23] Humans are mainly exposed to formaldehyde by inhalation from ambient and indoor air. Levels in indoor air may be much higher than in ambient air owing to evaporation of formaldehyde from furniture, painting and building constructions. Levels between 10 and $1000\,\mu g\,m^{-3}$ have been reported. The contribution from various atmospheric environments to the average human daily intake has been calculated to be $0.02\,mg\,d^{-1}$ for outdoor air, 0.5–$2\,mg\,d^{-1}$ for indoor conventional buildings and up to 1–$10\,mg\,d^{-1}$ for buildings with sources of formaldehyde. Smoking 20 cigarettes per day contributes an additional exposure of about 1 mg formaldehyde.[22]

Toxicokinetics

Formaldehyde is almost completely retained after inhalation; however, no incease in blood formaldehyde has been measured after exposure of humans or experimental animals. This may be explained by the high reactivity of formaldehyde. After retention of the highly hydrophilic substance in the mucus layer of the upper respiratory tract, formaldehyde either rapidly reacts with macromolecules in the mucus layer or the surface structures of the epithelial cells or is rapidly metabolized to formic acid and carbon dioxide by formaldehyde dehydrogenase and other enzymes, and becomes a source for the one-carbon biosynthetic pathways. In the rat, retention of formaldehyde takes place primarily in the nasal cavities, while in primates, formaldehyde may pass deeper into the respiratory tract and be retained in the trachea and the proximal regions of the major bronchi.[22,23]

Toxicological Effects

Exposure to formaldehyde vapour causes sensory irritation of the mucous membranes of the eyes and the upper respiratory tract. In humans the threshold of irritation is considered to be $0.1\,mg\,m^{-3}$ for the average population, while hyperreactive persons may experience irritation at lower levels.[22,24]

Formaldehyde reacts readily with macromolecules and has produced genotoxic effects *in vitro* and *in vivo*. DNA damage and mutations have been observed in bacterial assays, and DNA single-strand breaks, chromosomal aberrations, sister chromatid exchanges and gene mutations have been seen in assays using human cells. Studies have shown DNA–protein cross links in the nasal mucosa of rodents and monkeys, and chromosomal anomalies in lung cells of rats.[23]

[24] WHO, *Formaldehyde*, in *WHO Air Quality Guidelines for Europe*, 2nd edn., World Health Organization, Regional Office for Europe, Copenhagen, 1998.

Carcinogenicity

Several studies have demonstrated that long-term exposure to formaldehyde vapour by inhalation produces squamous cell carcinomas of the nasal epithelium of rats and mice.[23] The rat was found particulary sensitive, showing a steep dose–response relationship for the development of tumours in the exposure range of 7–18 mg m^{-3}.[25,26] As an example, when 120 rats of each sex were exposed to 0, 2.5, 6.9 or 17.6 mg m^{-3} (6 h per day, 5 days per week for up to 24 months), 2 of 235 animals at the medium exposure level (6.9 mg m^{-3}) and 103 of 232 animals at the high exposure level (17.6 mg m^{-3}) developed nasal squamous cell carcinomas.[25]

Much research has been done to clarify the mechanisms behind the carcinogenic effect of formaldehyde. It is recognized that nasal squamous cell carcinomas have only been observed at formaldehyde levels higher than those producing cytotoxic effects in the same tissue. In the rat, formaldehyde has induced cytotoxic responses and increased cell proliferation at exposure levels down to 2.5 mg m^{-3}.[25,27] Thus, the development of squamous cell carcinomas at higher exposure levels in animal experiments is considered a consequence of prolonged, repeated exposure to cytotoxic formaldehyde levels that produce inflammation, cell damage and increased cell proliferation.[23,28,29]

Epidemiological Data

More than 30 epidemiological studies have been performed on the relationship between exposure to formaldehyde and cancer in humans. From two meta-analyses a causal relationship was suggested for occupational exposure to formaldehyde and elevated risks for development of sinonasal and nasopharyngeal cancers.[30,31] On the basis of the overall human database, the IARC in 1995 concluded that there was only limited evidence for carcinogenic effects of formaldehyde in humans. However, the suggested relationship between formaldehyde exposure and cancer of the nasopharynx and the nasal cavities was noticed.[23]

Recent epidemiological data add further support to a causal relationship between formaldehyde exposure and nasal cancer. Among 3304 cancer patients having worked more than 10 years in 265 identified companies where formaldehyde was used, the only increased cancer risk was found for nasal cancer [RR for men: 2.3 (1.3–4.0, 95% confidence limit); RR for women: 2.4 (0.6–6.0, 95% confidence limit)]. In a subgroup of workers with no probable combined exposure to formaldehyde and wood-dust (a major confounder for nasal cancer), a relative risk for nasal cancer of 3.0 (1.4–5.7) was obtained.[32]

[25] W. D. Kerns, K. L. Pavkov, D. J. Donofrio, E. J. Gralla and J. A. Swenberg, *Cancer Res.*, 1983, **43**, 4382.

[26] T. M. Monticello, J. A. Swenberg, E. A. Gross, J. R. Leininger, J. S. Kimbell, S. Seilkop, T. B. Starr, J. E. Gibson and K. T. Morgan, *Cancer Res.*, 1996, **56**, 1012.

[27] E. Roemer, H. J. Anton and R. Kindt, *J. Appl. Toxicol.*, 1993, **13**, 103.

[28] H. A. Heck, M. Casanova and T. B. Starr, *Crit. Rev. Toxicol.*, 1990, **20**, 397.

[29] K. T. Morgan, *Toxicol. Pathol.*, 1997, **25**, 291.

[30] A. Blair, R. Saracci, P. A. Stewart, R. B. Hayes and C. Shy, *Scand. J. Work Environ. Health*, 1990, **16**, 381.

[31] T. Partanen, *Scand. J. Work Environ. Health*, 1993, **19**, 8.

[32] J. Hansen and J. H. Olsen, *Ugeskrift for Læger*, 1996, **158**, 4191 (in Danish with English summary).

Evaluation and Risk Assessment

The IARC has evaluated formaldehyde as probably carcinogenic to humans (group 2A), based on sufficient evidence for carcinogenicity from experimental animal data and limited evidence for carcinogenicity from human data.[23]

Based on data from a rat study[25] and using the linearized multistage model for cancer risk extrapolation, the US EPA in 1987 estimated a human inhalation unit risk of 1.6×10^{-2} per ppm of formaldehyde (corresponding to a unit risk of 1.3×10^{-5} per μg m^{-3}). In 1991, the US EPA re-evaluated this risk estimate in order to account for differences between rodents and primates in the anatomical structure of the nasal cavities. Using data from tissue dosimetry (formaldehyde-induced formation of DNA–protein cross links in target tissue of rats and monkeys), a human unit risk value of 3.3×10^{-4} per ppm was calculated (corresponding to a unit risk of 2.75×10^{-7} per μg m^{-3}). This was an overall 50-fold reduction of the previous risk estimate.[33,34] With this estimate an average level of formaldehyde in urban air of $10\,\mu$g m^{-3} corresponds to 2–3 extra cases of nasal cancer among 1 million exposed during a lifetime.

The WHO, in its updating of the Air Quality Guidelines for Europe, did not recommend any unit risk level for the carcinogenicity of formaldehyde. It was concluded that formaldehyde exposure below the cytotoxic level may only represent a negligible cancer risk, as cytotoxicity and repeated cell damage were considered closely linked to the development of nasal cancer. Thus, a guideline value of $100\,\mu$g m^{-3} to protect against sensory irritation was considered adequate to protect also against carcinogenic effects.

5 Other Aldehydes

Acetaldehyde is another aldehyde that has shown carcinogenicity in experimental animals. Acetaldehyde induced tumours in the nasal cavities in rats after long-term exposure by inhalation to 1350 mg m^{-3}, *i.e.* at much higher levels than formaldehyde. The lowest observed effect level for cytotoxicity in the nasal mucosa of rats was reported to be 275 mg m^{-3}; thus acetaldehyde may be considered much less potent than formaldehyde. Levels of acetaldehyde in urban ambient air average about $5\,\mu$g m^{-3}.[35] Based on this, the carcinogenic potential of acetaldehyde in ambient air seems far less problematic compared to formaldehyde.

Acrolein has higher potency than formaldehyde with respect to irritative effect and cytotoxicity. No carcinogenicity testing has been performed with acrolein. Acrolein has shown genotoxicity *in vitro*. The reported levels in ambient air tend to be somewhat lower than those of formaldehyde (about three times).[36]

[33] US EPA, *Formaldehyde Risk Assessment Update*, Office of Toxic Substances, US Environmental Protection Agency, Washington, 1991.

[34] O. Hernandez, L. Rhomberg, K. Hogan, C. Siegel-Scott, D. Lai, G. Grindstaff, M. Henry and J. A. Cotruvo, *J. Hazard Mater.*, 1994, **39**, 161.

[35] IPCS, *Acetaldehyde*, Environmental Health Criteria 167, International Programme on Chemical Safety, World Health Organization, Geneva, 1995.

[36] IPCS, *Acrolein*, Environmental Health Criteria 127, International Programme on Chemical Safety, World Health Organization, Geneva, 1992.

6 Polynuclear Aromatic Hydrocarbons (PAHs)

Sources and Levels in Ambient Air

PAHs are formed by incomplete combustion of organic materials and in natural processes such as carbonization. Major sources are residential heating (coal, wood, oil), automobile exhaust, industrial power generation, incinerators, the production of coal tar, coke, asphalt and petroleum catalytic cracking, cooking and tobacco smoking. Of the many hundreds of PAHs, the best known is benzo[*a*]pyrene (BaP), which often is used as a marker for PAHs in ambient air. In addition to the 'classical' PAHs, a number of heterocyclic aromatic compounds (*e.g.* carbazole, acridine), as well as derivatives of PAHs, such as nitro-PAHs and oxygenated PAHs, can be generated by incomplete combustion and from chemical reactions in ambient air. The contributions from the different important sources are difficult to estimate, and may vary from country to country. Stationary sources account for a high percentage of the total annual PAH emission. However, in urban or suburban areas, mobile sources are additional major contributors to PAH releases to the atmosphere. In air, PAHs are mainly attached to particles.[37][39] In the 1960s the annual average concentration of BaP was reported to be higher than 100 ng m^{-3} in several European cities.[37] In most developed countries, PAH concentrations have decreased substantially during the last 30 years owing to improved combustion technologies, increased use of catalytic converters in motor vehicles, and replacement of coal with oil and natural gas as energy sources. BaP levels below 1 ng m^{-3} are normally found at rural sites, while the levels in urban areas and areas with heavy traffic generally are 1–10 ng m^{-3}. PAH levels are higher during winter than summer.[39,40] In Copenhagen, the mean BaP concentration (January to March 1992) at a petrol station in a busy street was found to be 4.4 ng m^{-3}.[41]

Human Exposure

Human exposure to PAHs from inhalation of ambient air varies according to the degree of urbanization, traffic and industrialization. In terms of daily BaP intake the range could normally cover from less than 10 ng to more than 100 ng. Additional contributions from tobacco smoking and the use of unvented heating sources can increase PAH concentrations in indoor air, in certain cases to very high levels. Very high concentrations of PAHs can also occur in workplaces, such as coke-oven batteries, retort houses of coal-gas works and in the metal smelting industry.[37,42]

[37] WHO, *Polynuclear Aromatic Hydrocarbons (PAH)*, in *Who Air Quality Guidelines for Europe*, 2nd edn., World Health Organization, Regional Office for Europe, Copenhagen, 1998.

[38] ATSDR, *Toxicological Profile for Polycyclic Aromatic Hydrocarbons (PAHs): Update*, US Department of Health and Human Services, Public Health Services, Agency for Toxic Substances and Disease Registry, Atlanta, 1994.

[39] S. O. Baek, R. A. Field, M. E. Goldstone, P. W. Kirk, J. N. Lester and R. Perry, *Water Air Soil Pollut.*, 1991, **60**, 279.

[40] H. U. Pfeffer, *Sci. Total Environ.*, 1994, **146/147**, 263.

[41] T. Nielsen, H. E. Jørgensen, M. Poulsen, F. Palmgren Jensen, J. C. Larsen, M. Poulsen, A. B. Jensen, J. Schramm and J. Tønnesen, *Traffic PAH and Other Mutagens in Air in Denmark*, Miljøprojekt 285, Danish Environmental Protection Agency, Copenhagen, Denmark, 1995.

However, food is considered the major source of human exposure to PAH owing to the formation of PAH during cooking and from atmospheric deposition of PAHs on grains, fruits and vegetables. When human exposure to eight carcinogenic PAHs (benz[*a*]anthracene, chrysene, benzo[*b*]fluoranthene, benzo[*k*] fluoranthene, BaP, indeno[1,2,3-*cd*]pyrene, dibenz[*a,h*]anthracene and benzo[*ghi*] perylene) was estimated for a 'reference man', a mean total intake of 3.12 μg d^{-1} was estimated for non-smokers, of which food contributed 96.2%, air 1.6%, water 0.2% and soil 0.4%. Smokers consuming one pack of non-filtered cigarettes per day had an estimated additional intake of 1–5 μg d^{-1}.[43]

Toxicokinetics

PAHs are highly lipid-soluble and are absorbed from the lung, gut and skin. Studies in rats given microcrystalline PAHs or PAHs in solution have indicated that PAHs are efficiently cleared from the respiratory tract. For several PAHs, greater than 85% of the initial dose was cleared with a half-time of less than 1 hour.[37] Prolonged retention and extensive metabolism and activation of BaP was found to take place in the tracheobronchial epithelium of dogs.[44] When BaP is adsorbed on particles, the respiratory uptake rate is much lower, but the particles are retained for a long period of time in the respiratory tract. When radioactively labelled BaP adsorbed onto diesel engine exhaust particles or urban air particles was inhaled by rats, the lung clearance of the inhaled particle-associated radioactivity occurred in two phases: an initial rapid clearance and a long-term component that represented 50% of the radioactivity initially deposited in the lungs. Inhalation of [^{14}C]BaP adsorbed on carbon black particles resulted in 100-fold higher levels of ^{14}C in lungs at the end of a 12-week exposure than did inhalation of pure BaP.[37]

Irrespective of the route of administration, PAHs are rapidly and widely distributed throughout the organism. The highest levels are obtained in the liver.[45,46] Owing to the rapid metabolism of PAHs, no significant accumulation takes place in body fat. Studies show that BaP can readily cross the placental barrier of rats and mice. Following metabolism, hepatobiliary excretion and elimination through the faeces, and to a minor extent urinary excretion, are the major routes by which BaP metabolites are removed from the body, independent of the route of administration.[37,46-48]

[42] P. L. Lioy, J. M. Waldman, A. Greenberg, R. Harkov and C. Pietarinen, *Arch. Environ. Health*, 1988, **43**, 304.

[43] C. A. Menzie, B. B. Potocki and J. Santodonato, *Environ. Sci. Technol.*, 1992, **26**, 1278.

[44] P. Gerde, B. A. Muggenburg, J. R. Thornton-Manning, J. L. Lewis, K. H. Pyon and A. R. Dahl, *Carcinogenesis*, 1997, **18**, 1825.

[45] IARC, *Polynuclear Aromatic Compounds. Part 1. Chemical, Environmental and Experimental Data*, IARC Monographs on the Evaluation of the Carcinogenic Risk of Chemicals to Humans, vol. 32, International Agency for Research on Cancer, Lyon, France, 1983.

[46] H. Foth, R. Kahl and G. F. Kahl, *Food Chem. Toxicol.*, 1988, **26**, 45.

[47] J. R. Withey, J. Shedden, F. C. Law and S. Abedini, *J. Appl. Toxicol.*, 1993, **13**, 193.

[48] J. A. van de Wiel, P. H. Fijneman, C. M. Duijf, R. B. Anzion, J. L. Theuws and R. P. Bos, *Toxicology*, 1993, **80**, 103.

Metabolism and Activation

The biotransformation and activation of BaP and other PAHs have been studied intensively.[45,49-52] PAHs are metabolized by the microsomal cytochrome P450 system, which converts the non-polar PAHs into polar hydroxy and epoxy derivatives. Isozymes belonging to the P450 1A (in particular), P450 2A, P450 3A and P450 2B superfamilies are the major forms thought to be involved in the metabolism and activation of PAHs. These enzymes are widely distributed in cells and tissues of humans and animals. The highest metabolizing capacity is present in the liver, followed by the lung, intestinal mucosa, skin and kidneys.[37] In human lung, P450 1A1 seems to be the only inducible form of the P450 1As present. The P450 1A related activities in human lung are approximately 1–4% of those in the liver.[5]

BaP is initially oxidized to several arene oxides and phenols. The arene oxides may rearrange spontaneously to phenols, undergo hydration (catalysed by microsomal epoxide hydrolases) to the corresponding *trans*-dihydrodiols, or may react covalently with glutathione, either spontaneously or catalysed by cytosolic glutathione *S*-transferases. The phenols can be further oxidized to quinones. The phenols, quinones and dihydrodiols can all be conjugated to form glucuronides and sulfate esters, and the quinones also form glutathione conjugates. In addition, secondary epoxides are derived from the phenols and the dihydrodiols (resulting in diol epoxides) following further oxidation by the cytochrome P450 system.[45,53]

BaP and other PAHs stimulate their own metabolism by inducing cytochrome P450s and epoxide hydrolases. The induction of P450 1A1 is mediated by binding to a cytosolic receptor protein, the Ah receptor.[54] Numerous studies have shown that this induction leads to an enhanced turn-over of PAHs and enhanced generation of the active metabolites that bind to cellular macromolecules.[50,55]

Mechanism of Action

PAHs exert their mutagenic and carcinogenic activity through biotransformation to chemically reactive intermediates which bind covalently to cellular macromolecules (*i.e.* DNA).[45,49-53,56-59] Systematic studies indicate that vicinal

[49] W. Levin, A. Wood, R. Chang, D. Ryan and P. Thomas, *Drug Metab. Rev.*, 1982, **13**, 555.

[50] A. H. Conney, *Cancer Res.*, 1982, **42**, 4875.

[51] P. L. Grover, *Xenobiotica*, 1986, **16**, 915.

[52] A. Dipple, in *DNA Adducts: Identification and Biological Significance*, ed. K. Hemminki, A. Dipple, D. E. G. Shuker, F. F. Kadlubar, D. Segerbäck and H. Bartsch, IARC Scientific Publications 125, International Agency for Research on Cancer, Lyon, 1994, p. 107.

[53] A. Graslund and B. Jernstrom, *Q. Rev. Biophys.*, 1989, **22**, 1.

[54] D. W. Nebert, A. Puga and V. Vasiliou, *Ann. N. Y. Acad. Sci.*, 1993, **685**, 624.

[55] C. S. Cooper, P. L. Grover and P. Sims, in *Progress in Drug Metabolism*, ed. J. W. Bridges and L. F. Chasseaud, Wiley, Chichester, 1983, vol. 7, p. 295.

[56] D. H. Phillips and P. L. Grover, *Drug Metab. Rev.*, 1994, **26**, 443.

[57] F. A. Beland and M. M. Marques, in *DNA Adducts: Identification and Biological Significance*, ed. K. Hemminki, A. Dipple, D. E. G. Shuker, F. F. Kadlubar, D. Segerbäck and H. Bartsch, IARC Scientific Publications 125, International Agency for Research on Cancer, Lyon, France, 1994, p. 229.

[58] G. R. Shaw and D. W. Connell, in *Reviews of Environmental Contamination and Toxicology*, ed. G. W. Ware, Springer, New York, 1994, vol. 135, p. 1.

(bay-region) diol epoxides are the ultimate mutagenic and carcinogenic species of alternant PAHs. The BaP-7,8-dihydrodiol is oxidized to the bay-region BaP-7,8-dihydrodiol 9,10-epoxide (BPDE), which binds (*via* the C-10 position) predominantly to the exocyclic amino group of guanine (N^2), and the N^2-guanine adduct seems to be the most important mutagenic lesion in the case of BaP.[55,56]

Non-alternant PAHs, such as the benzofluoranthenes and indeno[1,2,3-*cd*] pyrene, differ in their metabolic activation from the alternant PAHs in exerting their genotoxic effects through reactive metabolites others than simple diol epoxides.[56] The mononitro-PAHs, by oxidative metabolism, form metabolites similar to the products formed from their parent PAH, but in addition undergo reductive metabolism at the nitro group to *N*-hydroxylamines which yield reactive, DNA-binding species. Dinitropyrenes do not appear to be activated through oxidative pathways but rather by reduction of one of the nitro groups. The *N*-hydroxylamines from nitro-PAH predominantly form DNA adducts by binding to C^8 of guanine.[57] PAHs can also form free radicals owing to the action of peroxidating enzyme systems in several different tissues. Free radicals of PAHs, including BaP, bind to C^8 and N^7 in guanine and N^7 of adenine in DNA, leaving apurinic sites in DNA.[60] However, the significance of this pathway for PAH carcinogenicity is unclear.[61] The structures of stereoisomeric covalent PAH DNA adducts have recently been reviewed.[62]

Use of Biomarkers

A number of methods have been used for monitoring of human exposure to PAHs (biomarkers of exposure and effect).[63] Plasma levels of BaP and metabolites were reported higher in humans living in an urban–industrial area than in outer suburban subjects. Smokers also had higher levels than non-smokers.[64] Urinary 1-hydroxypyrene finds increasing use as a marker of exposure to PAHs. Pyrene is normally abundant in complex PAH mixtures, and elevated urinary levels of the metabolite 1-hydroxypyrene have been found in smokers, patients cutaneously treated with coal tar, workers exposed to creosote oil, coal tar distillery workers, road paving workers, coke-oven workers, and workers exposed to bitumen fumes. Significant correlations were obtained between urinary 1-hydroxypyrene of coke-oven workers and city residents and levels of pyrene and BaP in the ambient air.[64–66] In studies on PAH exposure of the

[59] W. E. Bechtold, J. D. Sun, R. K. Wolff, W. C. Griffith, J. Kilmer and J. A. Bond, *J. Appl. Toxicol.*, 1991, **11**, 115.

[60] E. L. Cavalieri and E. G. Rogan, *Xenobiotica*, 1996, **25**, 677.

[61] A. P. Reddy, D. Pruess Schwartz, C. Ji, P. Gorycki and L. J. Marnett, *Chem. Res. Toxicol.*, 1992, **5**, 26.

[62] N. E. Geacintov, M. Cosman, B. E. Hingerty, S. Amin, S. Broyde and D. J. Patel, *Chem. Res. Toxicol.*, 1997, **10**, 111.

[63] G. Talaska, P. Underwood, A. Maier, J. Lewtas, N. Rothman and M. Jaeger, *Environ. Health Perspect.*, 1996, **104**, 901.

[64] J. Larden, *Toxicology*, 1995, **101**, 11.

[65] B. E. Moen, R. Nilsson, R. Nordlinder, S. Øvrebø, K. Bleie, A. H. Skorve and B. E. Hollund, *Occup. Environ. Med.*, 1996, **53**, 692.

[66] P. Strickland, D. Kang and P. Sithisarankul, *Environ. Health Perspect.*, 1996, **104**, 927.

general population from air pollution, intake of PAHs from foods is a serious confounder.[67,68]

The ^{32}P-postlabelling assay is highly sensitive for detection and quantization of carcinogen–DNA adducts.[69,70] ^{32}P-Postlabelling analyses have been used to detect DNA adducts in the skin of mice and humans exposed to components of complex mixtures of PAHs, including coal-tar and fuel exhaust condensates.[37] The principal methods that have been developed for the detection of PAH adducts to white blood cell DNA and blood proteins (haemoglobin, albumin) have been reviewed.[69,71,72] Higher levels of PAH–DNA adducts in lymphocytes of occupationally exposed workers, bus drivers[73] and smokers have frequently been found. In several studies, significant correlations between the estimated PAH exposures and adduct levels were obtained, while in other studies no such correlations were found.[74] Large inter-individual variations in DNA adduct formation and persistence have been found in freshly isolated lymphocytes. Dietary sources, such as ingestion of charcoal-broiled beef, may greatly influence the level of PAH–DNA adducts in white blood cells, making it very difficult to monitor the effect of air pollution in the general population.[75,76] Another complicating factor is that the DNA adduct level only correlates to the PAH level at low to moderate exposures, while at high exposure levels, such as cigarette smoke or coke-oven fumes, saturation and non-linearity is observed, resulting in a lower DNA-binding potency.[77]

The effect of environmental pollution on DNA adducts in humans has been measured in a highly industrialized area of Poland. Local controls exhibited adduct levels and patterns similar to those of coke workers, while the levels in rural controls were 2–3 times lower. The results showed that the levels of aromatic adducts in white blood cell DNA did not linearly relate to ambient air levels of PAHs and that other sources such as food might be important contributors.[78] Seasonal variations, with much higher levels of DNA adducts in the winter time, were observed in all groups.[79]

[67] A. Vyskocil, Z. Fiala, D. Fialova, V. Krajak and C. Viau, *Hum. Exp. Toxicol.*, 1997, **16**, 589.

[68] P. Sithisarankul, P. Vineis, D. Kang, N. Rothman, N. Caporaso and P. Strickland, *Biomarkers*, 1997, **2**, 217.

[69] K. Hemminki, H. Autrup and A. Haugen, *Toxicology*, 1995, **101**, 41.

[70] D. H. Phillips, *Mutat. Res.*, 1997, **378**, 1.

[71] M. dell'Omo and R. R. Lauwerys, *Crit. Rev. Toxicol.*, 1993, **23**, 111.

[72] M. C. Poirier and A. Weston, *Environ. Health Perspect.*, 1996, **104**, 883.

[73] P. S. Nielsen, N. de Pater, H. Okkels and H. Autrup, *Carcinogenesis*, 1996, **17**, 1021.

[74] K. Yang, L. Airoldi, R. Pastorelli, J. Restano, M. Guanci and K. Hemminki, *Chem.-Biol. Interact.*, 1996, **101**, 127.

[75] N. Rothman, M. C. Porier, R. A. Haas, A. Correa Villasenor, P. Ford, J. A. Hansen, T. O'Toole and P. T. Strickland, *Environ. Health Perspect.*, 1993, **99**, 265.

[76] K. Hemminki, C. Dickey, S. Karlsson, D. Bell, Y. Hsu, W.-Y. Tsai, L. A. Mooney, K. Savela and F. Perera, *Carcinogenesis*, 1997, **18**, 345.

[77] J. Lewtas, D. Walsh, R. Williams and L. Dobias, *Mutat. Res.*, 1997, **378**, 51.

[78] K. Hemminki, E. Grzybowska, M. Chorazy, K. Twardowska Saucha, J. W. Sroczynski, K. L. Putman, K. Randerath, D. H. Phillips, A. Hewer, R. M. Santella *et al.*, *Carcinogenesis*, 1990, **11**, 1229.

[79] E. Grzybowska, K. Hemminki and M. Chorazy, *Environ. Health Perspect.*, 1993, **99**, 77.

Toxicological effects

Haematological effects, including bone marrow toxicity, have been observed in animals following oral exposure to high doses of PAHs.[80] Data on the reproductive toxicity of PAHs are scarce. BaP is a potent reproductive toxicant adversely affecting the reproductive performance of pregnant rats and mice. *In utero* exposure to BaP has also produced several serious effects in the progeny of mice, such as testicular atrophy and interstitial cell tumours, immunosuppression and tumour induction.[37] A number of PAHs have an immunosuppressive effect in mice.[81]

A large number of PAHs and nitro-PAHs, as well as emissions containing these compounds, have shown genotoxicity and mutagenicity *in vitro* and *in vivo*.[38,45,82] The Ames test using various strains of *Salmonella typhimurium* has been widely used to monitor the mutagenic activity of PAHs and other mutagens present in complex environmental mixtures. Standard reference materials have been developed for mutagenicity assays of various combustion-related complex environmental mixtures.[83]

Carcinogenicity

A number of PAHs produce tumours in experimental animals. BaP has been used for many years as a model compound in studies of chemical carcinogenesis. When administered by the oral route, BaP and several other PAHs have produced tumours of the forestomach, liver, lungs and mammary glands of rodents,[45,84] while mono- and dinitropyrenes produced pituitary and mammary gland tumours.[85] BaP and other PAHs produce liver and lung tumours within half a year following intraperitoneal or subcutaneous injection to newborn animals.[45,84,86–89] In addition, nitro-PAHs produce leukaemia and tumours of the mammary glands and colon.[82,90] The study of skin tumours after dermal application of PAHs to mice has provided much of the background for the initiation/promotion theory in chemical carcinogenesis.[37,45,84]

[80] V. Anselstetter and H. Heimpel, *Acta Haematol. Basel*, 1986, **76**, 217.

[81] K. L. J. White, H. H. Lysy and M. P. Holsapple, *Immunopharmacology*, 1985, **9**, 155.

[82] IARC, *Diesel and Gasoline Engine Exhaust and Some Nitroarenes*, IARC Monographs on the Evaluation of Carcinogenic Risks to Humans, vol. 46, International Agency for Research on Cancer, Lyon, France, 1989.

[83] T. J. Hughes, J. Lewtas and L. D. Claxton, *Mutat. Res.*, 1997, **391**, 243.

[84] IARC, *Certain Polycyclic Aromatic Hydrocarbons and Heterocyclic Compounds*, IARC Monographs on the Evaluation of the Carcinogenic Risk of Chemicals to Humans, vol. 3, International Agency for Research on Cancer, Lyon, France, 1973.

[85] K. Imaida, M. S. Lee, C. Y. Wang and C. M. King, *Carcinogenesis*, 1991, **12**, 1187.

[86] K. L. Platt, E. Pfeiffer, P. Petrovic, H. Friesel, D. Beermann, E. Hecker and F. Oesch, *Carcinogenesis*, 1990, **11**, 1721.

[87] W. F. J. Busby, E. K. Stevens, C. N. Martin, F. L. Chow and R. C. Garner, *Toxicol. Appl. Pharmacol.*, 1989, **99**, 555.

[88] W. F. J. Busby, E. K. Stevens, E. R. Kellenbach, J. Cornelisse and J. Lugtenburg, *Carcinogenesis*, 1988, **9**, 741.

[89] E. J. Lavoie, J. Braley, J. E. Rice and A. Rivenson, *Cancer Lett.*, 1987, **31**, 15.

[90] K. Imaida, C. Uneyama, H. Ogasawara, S. Hayashi, K. Fukuhara, N. Miyata and M. Takahashi, *Cancer Res.*, 1992, **52**, 1542.

Several studies have confirmed the lung carcinogenicity of single PAHs and nitro-PAHs after direct application into the respiratory tract of rats and hamsters.[37,45,84,91-97] BaP is the only PAH that has been tested following inhalation. After long-term inhalation of 10 mg BaP per m^3, cancer of the respiratory tract occurred in 35% of golden hamsters.[37,98,99]

Dibenzo[*a,h*]anthracene, dibenzo[*a,h*]pyrene, dibenzo[*a,l*]pyrene, BaP, benzo[*b*]fluoranthene, 5-methylchrysene, 7*H*-dibenzo-[*c,g*]carbazole, 6-nitrochrysene and the dinitropyrenes and dinitrofluoranthenes are the strongest carcinogenic PAHs in animal bioassays. The benzofluoranthenes are moderately carcinogenic, while benz[*a*]anthracene and chrysene are relatively weak carcinogens.[37]

Carcinogenicity of PAH-containing Emissions

The 4–7 ring PAH fraction of condensate from car exhaust (gasoline, diesel), domestic coal stove emissions and tobacco smoke contains almost all the carcinogenic potential of PAHs. This has been found in a series of important studies using skin painting, subcutaneous injection and intrapulmonary implantation of different fractions (see refs. 37, 41). It can be concluded from the skin painting tests of different condensates that BaP represents about 5–15% of the carcinogenic potency of the exhaust condensates from petrol-driven vehicles and coal-fired domestic stoves. When tested by lung implantation in the rat, BaP contributed a somewhat lower percentage of the total carcinogenicity.

In rats exposed by inhalation to coal tar/pitch condensation aerosol containing either 20 or 46 μg m^{-3} BaP, a dose-related lung carcinogenic effect was observed. The lifetime lung tumour risk for rats exposed to 1 μg m^{-3} BaP as a constituent of this complex mixture was calculated to be 2%. In comparison, the estimated unit lung cancer risk for BaP based on epidemiological data from coking plants was 7–9%. It was suggested that in the evaluation of the lung carcinogenicity of PAHs attached to inhaled fine particles, the likely enhancing properties of the inflammatory effects of particles on lung tissue should be considered.[100]

Epidemiological and experimental inhalation studies indicate that cigarette smoke contains about 100 times less BaP, and diesel exhaust about 1000 times less BaP, than the exhaust from coke ovens or heated tar pitch which yield the

[91] R. Wenzel Hartung, H. Brune, G. Grimmer, P. Germann, J. Timm and W. Wosniok, *Exp. Pathol.*, 1990, **40**, 221.

[92] T. Maeda, K. Izumi, H. Otsuka, Y. Manabe, T. Kinouchi and Y. Ohnishi, *J. Natl. Cancer Inst.*, 1986, **76**, 693.

[93] R. P. Deutsch Wenzel, H. Brune, G. Grimmer, G. Dettbarn and J. Misfield, *J. Natl. Cancer Inst.*, 1983, **71**, 539.

[94] R. P. Deutsch Wenzel, H. Brune and G. Grimmer, *Cancer Lett.*, 1983, **20**, 97.

[95] M. Iwagawa, T. Maeda, K. Izumi, H. Otsuka, K. Nishifuji, Y. Ohnishi and S. Aoki, *Carcinogenesis*, 1989, **10**, 1285.

[96] S. Sato, H. Ohgaki, S. Takayama, M. Ochiai, T. Tahira, Y. Ishizaka, M. Nagao and T. Sugimura, *Dev. Toxicol. Environ. Sci.*, 1986, **13**, 271.

[97] K. Horikawa, N. Sera, T. Otofuji, K. Murakami, H. Tokiwa, M. Iawagawa, K. Izumi and H. Otsuka, *Carcinogenesis*, 1991, **12**, 1003.

[98] J. Thyssen, J. Althoff, G. Kimmerie and U. Mohr, *J. Natl. Cancer Inst.*, 1981, **66**, 575.

[99] J. Pauluhn, J. Thyssen, J. Althoff, G. Kimmerle and U. Mohr, *Exp. Pathol.*, 1985, **28**, 31.

[100] U. Heinrich, M. Roller and F. Pott, *Toxicol. Lett.*, 1994, **72**, 155.

same lung tumour incidence.[101] When results from inhalation studies with rats exposed to diesel engine exhaust ($4\,mg\,m^{-3}$) were compared with results from studies with rats exposed to coal-oven flue gas mixed with pyrolysed pitch, it was concluded that diesel exhaust had more lung tumour promotional effect than PAH-enriched coal-oven flue gas, which in turn was a more complete carcinogen.[102] The high levels of diesel exhaust used in the animal bioassays produce overloading of the lungs with particulate matter, resulting in severe inflammatory changes, while the lungs of rats exposed to coal-oven flue gas mixed with pyrolysed pitch had much less severe inflammatory changes. This suggests that the lung tumour response seen in rat inhalation bioassays using high level exposures to diesel soot is predominantly a non-specific effect, and to a lesser extent related to the genotoxic substances (*e.g.* PAHs, nitro-PAHs) present in diesel particulates. In support, it has been shown that fine carbon black particles and titanium dioxide particles, almost completely free of organic substances, were able to produce tumours in the rat lung after chronic inhalation exposure with particle mass concentrations in the exposure atmosphere of $6\,mg\,m^{-3}$.[103,104] The inhalation of titanium dioxide and metallic iron was also shown to produce impaired pulmonary clearance and persistent inflammation in the lungs of rats.[105]

Epidemiological Data

Epidemiological studies have focused on occupational exposures to PAHs. Occupational exposure to PAH-containing emissions from coke production, coal gasification, aluminium production, iron and steel founding, coal tars and coal tar pitches, and soots have produced lung cancer in humans, and coal tars and coal tar pitches, non-refined mineral oils, shale oils and soots have produced human skin and scrotal cancers.[106-108] Although PAHs are believed to be the main cause of cancer from these sources, a number of other compounds are present, probably also contributing to the effect. A particularly high rate of lung cancer mortality was found in coke-oven workers. The increases in lung cancer

[101] F. Pott and U. Heinrich, in *Complex Mixtures and Cancer Risk*, ed. H. Vainio *et al.*, IARC Scientific Publications 104, International Agency for Research on Cancer, Lyon, France, 1990, p. 288.

[102] R. O. McClellan, *Annu. Rev. Pharmacol. Toxicol.*, 1987, **27**, 279.

[103] U. Mohr, S. Tanenaka and D. L. Dungworth, *Dev. Toxicol. Environ. Sci.*, 1986, **13**, 459.

[104] U. Heinrich, R. Fusht, S. Rittinghausen, O. Creutzenberg, B. Bellmann, W. Koch and K. Levsen, *Inhal. Toxicol.*, 1995, **7**, 533.

[105] D. B. Warheit, J. F. Hansen, I. S. Yuen, D. P. Kelly, S. I. Snajdr and M. A. Hartsky, *Toxicol. Appl. Pharmacol.*, 1997, **145**, 10.

[106] IARC, *Polynuclear Aromatic Compounds: Part 3, Industrial Exposures to Aluminium Production, Coal Gasification, Coke Production, and Iron Steel Founding*, IARC Monographs on the Evaluation of the Carcinogenic Risk of Chemicals to Humans, vol. 34, International Agency for Research on Cancer, Lyon, France, 1984.

[107] IARC, *Polynuclear Aromatic Compounds: Part 4, Bitumens, Coal Tars and Derived Products, Shale-oils and Soots*, IARC Monographs on the Evaluation of the Carcinogenic Risk of Chemicals to Humans, vol. 35, International Agency for Research on Cancer, Lyon, France, 1985.

[108] IARC, *Overall Evaluations of Carcinogenicity: An Updating of IARC Monographs Volumes 1 to 42*, IARC Monographs on the Evaluation of the Carcinogenic Risk of Chemicals to Human, suppl 7, International Agency for Research on Cancer, Lyon, 1987.

cases correlate closely with the time spent working on top of ovens, where an average BaP concentration of about $30\,\mu\mathrm{g\,m^{-3}}$ has been detected.[37,106]

The US EPA has summarized a number of older risk estimates of lung cancer deaths based on epidemiological data and BaP levels in ambient air and in workplace air. The estimated unit risks ranged from 7.7×10^{-5} to 98×10^{-5} per ng BaP per $\mathrm{m^3}$.[109]

Evaluation and Risk Assessment

Assessment of Single PAHs. All epidemiological studies available have involved exposure to complex mixtures of which PAHs are only constituents. Therefore, there are no epidemiological studies that can be used to estimate the risk from exposure to a single PAH. On the basis of animal experiments, risk assessments of BaP and attempts to derive relative potencies of individual PAHs (relative to BaP) have been performed several times with the purpose of summarizing the contributions from selected PAHs into a total BaP equivalent dose, assuming additivity in their carcinogenic effects.[37,110-113] The main problem in absolute risk assessment is the lack of adequate data from long-term studies with purified PAHs. The multidose long-term studies so far available for risk assessment are mostly of an older date and concern only BaP. For BaP, unit lifetime risks calculated from experiments with various routes of administration, and using various modelling, yield comparable results. The range of unit lifetime risks calculated from a number of selected BaP studies included in a meta-analysis was 1.1×10^{-3} to 4.8×10^{-3} (mean: 3×10^{-3}) per $\mu\mathrm{g\,m^{-3}}$. A linearized multistage model was used, and the risk estimates were converted to human risk.[114] A new two-year gavage study with BaP in rats has been performed at the RIVM in the Netherlands, but has not yet been finally published. In that study, BaP produced dose-related increases of liver and forestomach tumours. Interestingly, high DNA adduct levels were found in many tissues, including the lungs, but measures of enhanced cell proliferation (due to toxicity) were only increased in the target organs.[115] A preliminary estimate of a unit risk derived from this study would expect it to be in the same order of magnitude as those calculated from the earlier studies.

Based on the extensive database on carcinogenicity studies using various routes of administration, an (entirely subjective) estimate of relative potencies for a number of PAHs is given in Table 1.[41] Such estimates should be used with great caution, as studies on mixtures of individual PAHs have shown that they may interact metabolically in a number of ways, resulting in synergistic, additive

[109] EPA, *Review and Evaluation of the Evidence for Cancer Associated with Air Pollution*, EPA-450/5-83-006R, US Environmental Protection Agency, Arlington, VA, 1984.

[110] I. C. T. Nisbet and K. LaGoy, *Regul. Toxicol. Pharmacol.*, 1992, **16**, 290.

[111] P. J. Rugen, C. D. Stern and S. H. Lamm, *Regul. Toxicol. Pharmacol.*, 1989, **9**, 273.

[112] T. W. Thorslund and D. Farrar, *Development of Relative Potency Estimates for PAHs and Hydrocarbon Combustion Product Fractions Compared to Benzo[a]pyrene and their Use in Carcinogenic Risk Assessment*, EPA/600/R-92/134, Dept. Commerce, NTIS, Springfield, VA, 1990.

[113] D. Krewski, T. Thorslund and J. Withey, *Toxicol. Ind. Health*, 1989, **5**, 851.

[114] J. F. Collins, J. P. Brown, S. V. Dawson and M. A. Marty, *Regul. Toxicol. Pharmacol.*, 1991, **13**, 170.

[115] D. Kroese, personal communication, January 1998.

Table 1 Best estimates of carcinogenic potencies of various PAHs, relative to BaP[a]

Compound	Relative potency	Compound	Relative potency
Anthracene	0.0005	Benzo[*a*]pyrene	1
Fluorene	0.0005	Benzo[*e*]pyrene	0.002
Phenanthrene	0.0005	Dibenz[*a,h*]anthracene	1.1
Benz[*a*]anthracene	0.005	Anthanthrene	0.3
Chrysene	0.03	Benzo[*ghi*]perylene	0.02
Cyclopenteno[*cd*]pyrene	0.02	Dibenzo[*a,e*]pyrene	0.2
Fluoranthene	0.05	Dibenzo[*a,h*]pyrene	1
Pyrene	0.001	Dibenzo[*a,i*]pyrene	0.1
Benzo[*b*]fluoranthene	0.1	Dibenzo[*a,l*]pyrene	1
Benzo[*j*]fluoranthene	0.05	Indeno[1,2,3-*cd*]pyrene	0.1
Benzo[*k*]fluoranthene	0.05	Coronene	0.01
Benzo[*ghi*]fluoranthene	0.01		

[a]Based on the authors' compilation of carcinogenicity studies in experimental animals using oral, pulmonary and skin application of PAH.[41]

or antagonistic effects, and nothing definitive can be concluded on the resulting tumorigenic actions of individual PAHs in complex mixtures.[116]

Assessment of Complex Mixtures of PAHs. A number of PAHs, nitro-PAHs, heterocyclic-PAHs, and their derivatives, have been found to be carcinogenic in animal experiments, and many more, including their reaction products in ambient air, have shown genotoxic effects *in vitro*. Therefore, the combined carcinogenic risk of exposure to the complex mixtures present in ambient air presumably is much higher than can be expected from either BaP alone or the use of BaP equivalent doses of the PAH determined by chemical measurements. If the composition of all emissions were identical and similar to those in ambient air, the concentration of a single PAH would provide a good index of the carcinogenic potential of the total mixture. However, this is not the situation.

The WHO, in its update of the Air Quality Guidelines for Europe, noted that the variations of PAH profiles in workplaces were not so wide and the deviation from the mean was relatively low in ambient air. The WHO therefore, on a provisional basis, used BaP as an index for the carcinogenic potential of general PAH mixtures in ambient air, although the limitations of the approach were recognized.[37]

The WHO based its evaluation on the mouse skin painting studies showing that 5–15% of the total carcinogenic effect from PAH fractions of different exhaust condensates was due to BaP, and adopted the lung cancer unit risk of 6.2×10^{-4} per $1\ \mu\text{g m}^{-3}$ of benzene-soluble compounds in ambient air, calculated by the US EPA from exposure to coke-oven emissions.[117] Assuming 0.71% BaP

[116] G. K. Montizaan, P. G. N. Kramers, J. A. Janus and R. Posthumus, *Integrated Criteria Document PAHs: Effects of 10 Selected Compounds*, Appendix to RIVM report no. 758474011, National Institute of Public Health and Environmental Protection, RIVM, Bilthoven, The Netherlands, 1989.

[117] EPA, *Carcinogen Assessment of Coke Oven Emissions*, EPA-600/6-82-003F, US Environmental Protection Agency, Office of Health and Environmental Assessment, Washington, DC, 1984.

in benzene-soluble coke-oven emissions, a lifetime risk of respiratory cancer of 8.7×10^{-2} per $\mu g \, m^{-3}$ BaP, or 8.7×10^{-5} per $ng \, m^{-3}$ BaP, was calculated.[37] This risk assessment implies that about 9 per 100 000 exposed individuals may die from cancer of the respiratory tract as a result of spending a lifetime in ambient air containing an average level of 1 ng BaP per m^3.

In a Dutch Integrated Criteria Document on PAHs a unit risk of 6.58×10^{-2} per $\mu g \, m^{-3}$ BaP was estimated from studies of Chinese women in the Xuan Wei province using smoky coal for cooking.[116]

In urban areas, air pollution from motor vehicles, in particular diesel cars, becomes increasingly important. Quantitative estimates of lung cancer risks from exposure to diesel engine particulate emissions have been done using data from the chronic bioassays with rats. Dose was based upon the concentration of carbon particulate matter per unit lung surface area. The unit risk estimates varied from 1.0 to 4.6×10^{-5}, with a geometric mean of 1.7×10^{-5} per $\mu g \, m^{-3}$ of diesel exhaust particulate matter.[118] This approach has been seriously questioned, because the carcinogenic effect seen in the rat studies with diesel particles is mainly due to a mechanism for which there seems to be a threshold.[119] Moreover, it also appears that the rat has a unique sensitivity to the lung effects of poorly soluble, non-fibrous particles, in contrast to the mouse and primates.[120,121] An examination of epidemiological studies of workers highly exposed to airborne carbon black showed no satisfactory evidence for an association with increased lung cancer risk, which was inconsistent with the predictions from the rat bioassay data.[122]

Biological assays have been used for the determination of the potency of various complex mixtures relative to the potency of BaP or potencies calculated on the basis of human data. The database of useful human studies is rather limited, and only a few sources of PAH emissions can be evaluated by this approach, notably cigarette smoke condensates, coke-oven emissions and roofing tar emissions. The *Salmonella* mutagenicity assay, the mouse skin tumour initiation-promotion assay and the rat lung implant assay have been evaluated for use in potency estimations of complex mixtures from combustion sources. Linear correlations have been established between mouse skin initiation potency, bacterial mutagenic potency and potencies calculated from human exposure. These correlations have been used to estimate the human potency of other mixtures from combustion sources for which only mouse skin initiation data or mutagenicity data exist. A particularly good correlation was noted between the mouse skin data and the human data, and good correlations were obtained between mouse skin data and mutagenicity for a series of mixtures from the same source category. When mixtures of emissions from different sources were assayed the correlations decreased, but were still reasonable. Using this *comparative potency approach*, a unit risk of 1.3×10^{-4} per μg organic matter m^{-3} was

[118] W. E. Pepelko and C. Chen, *Regul. Toxicol. Pharmacol.*, 1993, **17**, 52.

[119] P. A. Valberg and A. Y. Watson, *Regul. Toxicol. Pharmacol.*, 1996, **24**, 30.

[120] J. L. Mauderly, D. A. Banas, W. C. Griffith, F. F. Hahn, R. F. Henderson and R. O. McCellan, *Fund. Appl. Toxicol.*, 1996, **30**, 233.

[121] K. J. Nikula, K. J. Avila, W. C. Griffith and J. L. Mauderly, *Fund. Appl. Toxicol.*, 1997, **37**, 37.

[122] P. A. Valberg and A. Y. Watson, *Regul. Toxicol. Pharmacol.*, 1996, **24**, 155.

Table 2 Estimated lifetime cancer risk from air pollution components using different approaches

Exposure indicator	Average level[125]	Unit risk for indicator	Estimated lifetime cancer risk
Diesel particles	$1\ \mathrm{g\,m^{-3}}$	$1.7 \times 10^{-5}\,(\mu\mathrm{g\,m^{-3}})^{-1}$ [c]	1.7×10^{-5}
Extractable organic material from diesel particles[a]	$0.07\,\mu\mathrm{g\,m^{-3}}$	$2.3 \times 10^{-4}\,(\mu\mathrm{g\,m^{-3}})^{-1}$ [d]	1.6×10^{-5}
Extractable organic material from soot[b]	$1\,\mu\mathrm{g\,m^{-3}}$	$1.3 \times 10^{-4}\,(\mu\mathrm{g\,m^{-3}})^{-1}$ [d]	13×10^{-5}
Benzo[*a*]pyrene	$0.7\,\mathrm{ng\,m^{-3}}$	$8.7 \times 10^{-5}\,(\mathrm{ng\,m^{-3}})^{-1}$ [e]	6.1×10^{-5}

[a]Assuming 7% extractable organic material.[125]
[b]Assuming 20% extractable organic material.[41]
[c]Derived from animal bioassays.[128]
[d]Derived by the comparative potency approach.[124]
[e]WHO Air Quality Guidelines for Europe.[37]

obtained for air particle extracts containing 64% contribution from woodsmoke and 36% from mobile source emissions. For automobile diesel (light duty) and automobile gasoline, unit risks of 2.3×10^{-4} and 1.1×10^{-4}, respectively, per μg organic matter m^{-3} were estimated.[123,124]

Using information on the average levels reported in Sweden[125] of diesel particles ($1\ \mu\mathrm{g\,m^{-3}}$), soot ($5\ \mu\mathrm{g\,m^{-3}}$) and BaP ($0.7\ \mathrm{ng\,m^{-3}}$), several risk estimates can be compared (Table 2). It should be noted that, despite the critique of using rat bioassay data in the risk assessment of diesel exhaust, the use of the comparative potency approach on diesel particles yields a similar risk estimate, probably by coincidence. Somewhat higher risk estimates are obtained when the approach recommended by the WHO[37] is used, and when the comparative potency approach is used on total soot. This is to be expected, since diesel exhaust contributes only a part of the total air pollution with PAHs. It can be added that when epidemiological data from exposures to diesel exhausts are used, an even higher unit risk (30×10^{-5}) has been estimated.[125]

7 Conclusion

The available epidemiological evidence indicates that combustion-related emissions are related to an increased risk of lung cancer from general air pollution. The increased risk appears to be small, compared to the risk from tobacco smoking. This evidence is supported by epidemiological studies from occupational settings and from toxicological studies *in vitro* and in experimental animals with single compounds and complex mixtures related to air pollution. In addition to an increased lung cancer risk from air pollution, compounds that have been found to be carcinogenic in occupational epidemiology and animal experiments are also

[123] J. Lewtas, *Environ. Health Perspect.*, 1993, **100**, 211.
[124] J. Lewtas, *Pharmacol. Toxicol.*, 1993, **72** (suppl. 1), s55.
[125] M. Törnqvist and L. Ehrenberg, *Environ. Health Perspect.*, 1994, **102** (suppl. 4), 173.

present, which may contribute to human cancer at other sites than the lungs, such as leukaemia and cancers of the lymphohaematopoietic system.

The major causal chemical factors that contribute to the cancer risk from air pollution are difficult to identify, because the vast number of chemicals that can be suspected on the basis of their toxicological properties and occupational epidemiology appear in the ambient air in extremely complex mixtures, the composition of which may vary to a large extent from one location to another. In addition to the large number of PAHs and derivatives, benzene, 1,3-butadiene and aldehydes, mentioned in this paper, many other compounds such as styrene, dichloromethane, tri- and tetrachloroethylenes, polychlorinated *p*-dibenzodioxins and dibenzofurans, polychlorinated biphenyls, a number of inorganics and radionuclides may also contribute to the overall cancer risk from air pollution.

The link between the epidemiological evidence for lung cancer risk and the occurrence of PAHs in ambient air is particularly strong, and is supported by data from animal studies, although a definite risk estimate is difficult to obtain. PAHs are locally deposited with particles in the lungs, are slowly released over a long period of time, undergo extensive metabolic activation in the lungs, and in general tend to produce tumours at the site of application, in this case the lungs. Keeping in mind that the carcinogenic process involves both genotoxic and non-genotoxic mechanisms, it is of particular importance for future research to elucidate to what extent the well-known genotoxic properties of PAHs (and other genotoxic agents) act in concert with the irritant and inflammatory properties of other factors in air pollution, such as ozone, aldehydes and particulate matter.

Setting Health-based Air Quality Standards

ROY M. HARRISON

1 Introduction

Health-based air quality standards form one of the cornerstones of the air quality management process. Progress in improving air quality without explicit air quality standards is possible, as occurred in the UK between the 1950s and 1980s, but in any modern rational system of air quality management, some form of health-based objective is needed, and generally these are referred to as air quality standards.

To set air quality standards in context it is necessary to understand the entire air quality management process (see Figure 1); the reader is referred to recent articles by Middleton[1] and Lloyd.[2] There are basically three major facets:

(a) *Monitoring and public information.* This provides the necessary knowledge of current air quality and can inform people of periods when some form of preventative or protective activity is desirable at the individual, corporate or community level. Provision of data to the public also serves a purpose in generating informed public pressure for better air quality.

(b) *Air quality standards.* These are benchmarks of acceptability against which the monitoring data may be judged. They can serve as long-term objectives which control strategies are designed to meet or, as in the case of European Limit Values, can provide firm legislative ceilings which, if exceeded, can result in legal action and the requirement for immediate remedial measures.

(c) *Control policy design and implementation.* If air quality standards represent the desired endpoint, then this is the means of achieving it. Normally some form of numerical model is used to determine the most cost-effective means of reducing emissions so as to meet air quality standards, and subsequently legal controls are applied which are designed to bring about the necessary

---

[1] D. R. Middleton, in *Air Quality Management*, Issues in Environmental Science & Technology, vol. 8, ed. R. E. Hester and R. M. Harrison, Royal Society of Chemistry, Cambridge, 1997.

[2] A. C. Lloyd, in *Air Quality Management*, Issues in Environmental Science & Technology, vol. 8, ed. R. E. Hester and R. M. Harrison, Royal Society of Chemistry, Cambridge, 1997.

Figure 1 Schematic of the air quality management process

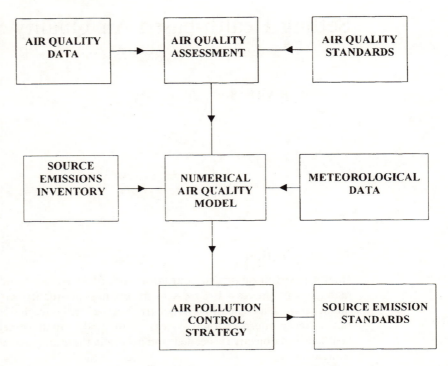

emissions reductions. The European Auto-Oil Programme described by Skouloudis[3] is an excellent example of this process in action.

Thus, air quality standards are the benchmark of acceptability of air quality. A more formal definition might be the following:

An air quality standard is a concentration of an air pollutant below which effects on human health are expected to be zero or negligibly small at a population level.

It must be recognized that air quality standards are designed to protect the health of populations rather than of every individual. Thus, highly susceptible individuals, *e.g.* brittle (very severe) asthmatics or those who contract cancer as a result of exposure to very low concentrations of environmental chemicals, may suffer serious personal consequences even when concentrations are within air quality guidelines. In general, air quality standards do seek to protect sensitive individuals such as normal asthmatics, but not the most sensitive members of the population; as will be discussed later, it is impracticable to set standards for genotoxic carcinogens which guarantee that there will be zero risk of consequent cancer cases.

There is no universally agreed distinction between the terms *standard*, *guideline* and *objective* when applied to air quality. In the UK the term 'standard' is used in

---

[3] A. N. Skouloudis, in *Air Quality Management*, Issues in Environmental Science & Technology, vol. 8, ed. R. E. Hester and R. M. Harrison, Royal Society of Chemistry, Cambridge, 1997.

Table 1 Air quality
standards recommended
by EPAQS

Pollutant	Concentration	Averaging period
Benzene	5 ppb	1 year running mean
1,3-Butadiene	1 ppb	1 year running mean
Sulfur dioxide	100 ppb	15 minutes
Nitrogen dioxide	150 ppb	1 hour
Ozone	50 ppb	8 hour running mean
Carbon monoxide	10 ppm	8 hour running mean
PM_{10}	50 $\mu g\,m^{-3}$	24 hour running mean

the sense defined above, whilst specific objectives are shorter-term policy objectives to be met *en route* to long-term compliance with the standard. The World Health Organization, however, produces what it refers to as guidelines which nonetheless conform to the above definition of a standard. The WHO uses this term in order to convey the idea that individual countries should set their own standards based on the WHO guidelines, but taking regard of individual socio-economic considerations. Thus, a national standard could be greater than or less than the corresponding WHO guideline.

2 Sources of Air Quality Standards

There are now a number of authoritative independent sources of air quality standards. This article will make no attempt to review exhaustively or intercompare such standards, and indeed, the majority of the discussion will centre around the work of the UK Expert Panel on Air Quality Standards (EPAQS) and the WHO Working Group which produced the revised WHO Air Quality Guidelines for Europe. A summary of the EPAQS recommendations[4-10] available at the time of writing is presented in Table 1 and a selection of the WHO recommendations,[11] omitting those for indoor air and ecotoxic effects, is presented in Table 2. Other sources of air quality standards are the European Commission and the US Environmental Protection Agency. These differ from the EPAQS and WHO recommendations in one important regard: both can have legal force.

The standards recommended by EPAQS (see Table 1) have been adopted by the UK Government as long-term benchmarks for air quality. In some instances they have been translated into Objectives in the National Air Quality Strategy, which take into account costs and benefits and which it is intended to achieve by the year 2005. Objectives for seven pollutants have been set down in regulation (The Air Quality Regulations 1997), which triggers a duty, set down in Part IV of

[4] Expert Panel on Air Quality Standards, *Benzene*, HMSO, London, 1994.
[5] Expert Panel on Air Quality Standards, *Ozone*, HMSO, London, 1994.
[6] Expert Panel on Air Quality Standards, *Carbon Monoxide*, HMSO, London, 1994.
[7] Expert Panel on Air Quality Standards, *1,3-Butadiene*, HMSO, London, 1994.
[8] Expert Panel on Air Quality Standards, *Sulphur Dioxide*, HMSO, London, 1995.
[9] Expert Panel on Air Quality Standards, *Particles*, HMSO, London, 1995.
[10] Expert Panel on Air Quality Standards, *Nitrogen Dioxide*, HMSO, London, 1996.
[11] World Health Organization, *Revised Air Quality Guidelines for Europe*, WHO European Office, Copenhagen, 1998.

Table 2 Summary of
objectives in the UK
National Air Quality
Strategy[12]

| Pollutant | Standard | | Objective—to be achieved by 2005 |
	concentration	measured as	
Benzene	5 ppb	Running annual mean	5 ppb
1,3-Butadiene	1 ppb	Running annual mean	1 ppb
Carbon monoxide	10 ppm	Running 8-hour mean	10 ppm
Lead	0.5 μg m^{-3}	Annual mean	0.5 μg m^{-3}
Nitrogen dioxide	150 ppb	1 hour mean	150 ppb, hourly mean[a]
	21 ppb	Annual mean	21 ppb, annual mean[a]
Ozone	50 ppb	Running 8-hour mean	50 ppb, measured as the 97th percentile[a]
Fine particles	50 μg m^{-3}	Running 24-hour mean	50 μg m^{-3} measured as the 99th percentile[a]
Sulfur dioxide	100 ppb	15 minute mean	100 ppb measured as the 99.9th percentile[a]

[a]These objectives are to be regarded as provisional.

the Environment Act 1995, upon local government to address issues of unacceptable air quality, as defined by the air quality objectives, within their boundaries. There are, however, no penalties for exceedence. The Standards and Objectives adopted in the March 1997 UK National Air Quality Strategy,[12] now under review, appear in Table 2. For some pollutants, such as nitrogen dioxide and benzene, the Objective is identical to the Standard. For others, however, a percentile compliance varying from 97% for ozone to 99.9% for sulfur dioxide is specified. Thus, some exceedences of the Standard are envisaged even in the year 2005.

The philosophy used by EPAQS to recommend standards has been very similar to that used by the WHO in setting their air quality guidelines for Europe. The first set of European guidelines was published in a book entitled 'Air Quality Guidelines for Europe' in 1987 and have proved to be extremely influential. The guidelines for the 'irritant' air pollutants, carbon monoxide and trace metals were designed to protect even sensitive members of the population and to incorporate an additional margin of safety, and this ethos is also fundamental to the recent revisions.[11] This is also the philosophy behind EPAQS standards. The major difference between WHO recommendations and EPAQS is that for the non-threshold pollutants, such as the genotoxic carcinogens and PM_{10}, the WHO cites exposure–response gradients rather than recommending a guideline. It is therefore left to individual governments to determine national standards

[12] Department of the Environment, *The United Kingdom National Air Quality Strategy*, Stationery Office, London, 1997.

Table 3 Revised WHO air quality guidelines for Europe (second edition)

Substance	Guideline value	Averaging time
Classical air pollutants		
Carbon monoxide	$100 \ mg \ m^{-3}$	15 min
	$60 \ mg \ m^{-3}$	30 min
	$30 \ mg \ m^{-3}$	1 hour
	$10 \ mg \ m^{-3}$	8 hour
Ozone	$120 \ \mu g \ m^{-3}$	8 hour
Nitrogen dioxide	$200 \ \mu g \ m^{-3}$	1 hour
	$40 \ \mu g \ m^{-3}$	Annual
Sulfur dioxide	$500 \ \mu g \ m^{-3}$	10 min
	$125 \ \mu g \ m^{-3}$	24 hour
	$50 \ \mu g \ m^{-3}$	Annual
Particulate matter	Exposure–response	
Inorganic pollutants		
Arsenic	$1.5 \times 10^{3} (\mu g \ m^{-3})^{-1}$	UR[a]/lifetime
Cadmium	$5 \ ng \ m^{-3}$	Annual
Chromium (Cr^{VI})	$4 \times 10^{-2} (\mu g \ m^{-3})^{-1}$	UR[a]/lifetime
Fluoride	No guideline	
Lead	$0.5 \ \mu g \ m^{-3}$	Annual
Manganese	$0.15 \ \mu g \ m^{-3}$	Annual
Mercury	$1.0 \ \mu g \ m^{-3}$	Annual
Nickel	$3.8 \times 10^{-4} (\mu g \ m^{-3})^{-1}$	UR[a]/lifetime
Platinum	No guideline	
Organic pollutants		
Benzene	$6 \times 10^{-6} (\mu g \ m^{-3})^{-1}$	UR[a]/lifetime
1,3-Butadiene	No guideline	
Dichloromethane	$3 \ mg \ m^{-3}$	24 hour
Formaldehyde	$0.1 \ mg \ m^{-3}$	30 min
PAH (BaP)	$8.7 \times 10^{-5} (ng \ m^{-3})^{-1}$	UR[a]/lifetime
Styrene	$0.26 \ mg \ m^{-3}$	1 week
Tetrachloroethylene	$0.25 \ mg \ m^{-3}$	24 hour
Toluene	$0.26 \ mg \ m^{-3}$	1 week
Trichloroethylene	$4.3 \times 10^{-7} (\mu g \ m^{-3})^{-1}$	UR[a]/lifetime

[a]UR = Unit Risk (see text).

suited to local circumstances. Further discussion of the unit risk value is provided later in this article. In the last year or two the WHO European office, using advisers drawn internationally, has revised the air quality guidelines for Europe, and at the time of writing a second edition is in the press. The guidelines listed in Table 3 are the outcome of that revision process.

In contrast, the European Union has limit and guide values for a number of pollutants. The limit values have legal force and unless a specific derogation is granted, member states are expected to ensure that air quality complies with the limit values embodied in the various air quality Directives. Currently, Directive limit values are in force for sulfur dioxide, smoke/particulate matter, nitrogen dioxide and lead, and public information thresholds and alert levels are set for ozone. The EU has, however, recently adopted a new framework Directive on

Table 4 Substances for which EU air quality standards are currently proposed	Sulfur dioxide Nitrogen dioxide Fine particulate matter Suspended particulate matter Lead Ozone as at present (Directive 92/72/EC) Benzene Carbon monoxide Polyaromatic hydrocarbons Cadmium Arsenic Nickel Mercury

Ambient Air Quality Assessment and Management (96/62/EC) under which individual Daughter Directives will be set, dealing with monitoring protocols and air quality standards for a range of specific pollutants. The current list of substances for which EU air quality standards are to be developed is given in Table 4. At the time of writing, formal proposals have been published for the first five substances on the list (fine and suspended particulate matter are taken together as PM), and these appear in Table 5. Working groups are currently developing further proposals on carbon monoxide, benzene and ozone. It must be emphasized that any proposed values may well be changed before entry into community law.

In the US, ambient air quality standards are set by the US EPA and have legal force, although, in general, policy is directed towards long-term achievement of air quality standards rather than immediate prosecution of infringements. The current US EPA standards appear in Table 6. These differ somewhat from the EPAQS and WHO standards, at least in part because of the extensive consultation of industry and other pressure groups which goes on whilst setting the standards. The economic costs and benefits of compliance are not considered directly in the standards setting process, but an economic analysis is also conducted by the EPA (see chapter by M. Lippmann in this volume). In the UK, consideration of costs and benefits and consultation in the industry and pressure groups takes place when the Government publishes its response to the recommendations of EPAQS, and any draft national objectives that may be developed from those standards.

3 Setting Air Quality Standards

The essence of the process of setting air quality standards is easily explained. Maynard[13] describes the sequence in the following way:

(a) Understand the exposure–response relationship of the pollutant in question
(b) Decide on an acceptable level of effects

[13] R.L. Maynard, in *Setting and Managing Standards for Air Quality*, Cambridge Environmental Initiative Professional Seminar Series, Cambridge, 1994.

Table 5 Health-based limit values proposed by the European Commission

Pollutant	Limit value	Margin of tolerance	Target date
SO_2	$350 \, \mu g \, m^{-3}$ (1-hour average) not to be exceeded more than 24 times a year	43% on commencement of Directive, falling linearly to 0% between 1.1.2001 and 1.1.2005	1 January 2005
	$125 \, \mu g \, m^{-3}$ (24-hour average) not to be exceeded more than 3 times a year	None	1 January 2005
NO_2	$200 \, \mu g \, m^{-3}$ (1-hour average) not to be exceeded more than 8 times a year	50% on commencement of Directive, falling linearly to 0% between 1.1.2001 and 1.1.2010	1 January 2010
	$40 \, \mu g \, m^{-3}$ (annual average)	50% on commencement of Directive, falling linearly to 0% between 1.1.2001 and 1.1.2010	1 January 2010
PM_{10} (Stage 1)	$50 \, \mu g \, m^{-3}$ (24-hour average) not to be exceeded more than 25 times a year	50% on commencement of Directive, falling linearly to 0% between 1.1.2001 and 1.1.2005	1 January 2005
	$30 \, \mu g \, m^{-3}$ (annual average)	50% on commencement of Directive, falling linearly to 0% between 1.1.2001 and 1.1.2005	1 January 2005
PM_{10} (Stage 2)	$50 \, \mu g \, m^{-3}$ (24-hour average) not to be exceeded more than 7 times a year	None	1 January 2010
	$20 \, \mu g \, m^{-3}$ (annual average)	None	1 January 2010
Lead	$0.5 \, \mu g \, m^{-3}$ (annual average)	100% on commencement of Directive, falling linearly to 0% between 1.1.2001 and 1.1.2005	1 January 2005

Table 6 US ambient air quality standards

Pollutant	Measurement period	Concentration
Sulfur dioxide	Annual arithmetic mean 24 hour average (not to be exceeded more than once per year)	30 ppb ($80\,\mu g\,m^{-3}$) 140 ppb ($365\,\mu g\,m^{-3}$)
Particulate: PM_{10}	24 hour average (99%ile)	$150\,\mu g\,m^{-3}$
PM_{10}	Annual arithmetic mean	$50\,\mu g\,m^{-3}$
$PM_{2.5}$	24 hour average (98%ile)	$65\,\mu g\,m^{-3}$
$PM_{2.5}$	Annual arithmetic mean	$15\,\mu g\,m^{-3}$
Carbon monoxide	8 hour average (not to be exceeded more than once per year)	9 ppm ($10\,000\,\mu g\,m^{-3}$)
	1 hour average (not to be exceeded more than once per year)	35 ppm ($40\,000\,\mu g\,m^{-3}$)
Ozone	8 hour average (annual fourth highest daily maximum)	80 ppb
	1 hour average (not to be exceeded more than once per year)	120 ppb ($235\,\mu g\,m^{-3}$)
Nitrogen dioxide	Annual arithmetic mean	53 ppb (100s $\mu g\,m^{-3}$)
Lead	Maximum arithmetic mean averaged over a calendar quarter	$1.5\,\mu g\,m^{-3}$

(c) Set standard so that effects do not exceed those specified as acceptable

Clearly this process contains elements which will be interpreted differently by different standard setting bodies. The understanding of the exposure–response relationships has notably varied between standard setting bodies for some pollutants. This is usually because different degrees of weight are attached to the importance of the various published studies. Unfortunately, as will be described shortly, the base of data upon which exposure–response relationships may be established is often weak and sometimes contradictory, and hence different views may be taken according to the weight put on particular studies. For example, in the case of nitrogen dioxide, whilst population-based epidemiological studies have demonstrated effects at rather modest levels of exposure to NO_2, controlled exposure studies in the laboratory have shown that much higher concentrations of NO_2 are needed to elicit a response in these circumstances. There are various possible interpretations of this apparent contradiction, amongst them being that the chamber studies have been poorly designed and have failed to recognize important health outcomes, or, on the other hand, the population-based epidemiological studies have been subject to confounding from the co-variation of nitrogen dioxide with other pollutants such as particulate matter, and therefore the effects which they attribute to nitrogen dioxide are in fact the result of exposures to other pollutants. This point is elaborated on in the next section.

Furthermore, even if there is a clear understanding of the exposure–response relationships, then different panels will take a different view of what is an acceptable level of effects. Maynard[13] comments that 'on an international scale, little agreement on what constitutes acceptable effects has been reached: opinions vary from no effect to effects significantly less than those produced by other more uncontrollable environmental factors such as variations in temperature and epidemics of mild infections such as colds'. The acceptability will, in the eyes of some, also take account of costs as well as benefits and clearly the judgements made are highly individual and subjective overall.

4 Understanding Exposure–Response Relationships

The base of data available for understanding exposure–response relationships comes from four major sources. These are essentially complementary and, in an ideal world, information would be available from all four, and when integrated would give a coherent whole. Often this is not the case.

Controlled human exposure studies in the laboratory have the advantage of offering good definition of acute effects and precise knowledge of exposure concentrations. They are excellent for allowing identification of the effects of a single pollutant, and particularly good for pollutants which act as respiratory tract irritants and therefore elicit changes in lung function over relatively short time periods which are capable of being measured. Thus, for a small range of pollutants eliciting modest reversible effects, chamber studies provide excellent data which have proved useful in setting standards for pollutants such as nitrogen dioxide and sulfur dioxide. Whilst it should be possible to study simple mixtures in chambers, rather little work has been conducted with mixtures, and indeed, it is not possible to simulate the full complexity of an urban air pollutant mix. The measures of health impact used in chamber studies are generally tests of lung function such as FEV_1 (Forced Expiratory Volume in 1 second: the volume of air expired forcibly in one second) or FVC (Forced Vital Capacity: the total volume of air which can be expelled forcibly), but the health significance of small changes in these parameters when reversible, either for the individual or for the population as a whole, is extremely difficult to judge. Additionally, biochemical tests are now available which can demonstrate the initiation of inflammatory processes, but again the long-term health significance of such inflammation for the individual is very difficult to determine, and hence for studies showing an exposure–response gradient, it is very hard to know at what point on that curve to determine an acceptable level of effect. A useful application of chamber studies has been in research on the interaction between gaseous air pollutants and airborne allergens (*e.g.* grass pollen or house dust mite allergen), a topic very hard to study through epidemiology.

Secondly, there are *epidemiological studies in the general population*. These have the great advantage of offering definitions of effects of real pollutant mixtures on whole populations if confounding factors can be adequately controlled. The latter can be a major problem, since such apparently routine variables as air temperature can have major impacts on such profound outcomes as mortality. Nonetheless, epidemiological studies have tended to be the backbone of the standard setting process, proving extremely valuable for irritant pollutants and particles. Studies in which day-to-day changes in airborne pollutant loading are related to health service data such as hospital admissions have been especially persuasive. However, there are persistent doubts in some cases whether epidemiology can correctly identify the harmful pollutant in a complex urban mixture, or indeed whether it is the mix itself which is responsible for the observed response. Many of the important recent epidemiological studies have used a time series design. These involve relating measurements of daily average air pollution collected over a period of a year or more to a health outcome such as mortality or hospital admission on the same day or lagged by anything up to three days after the air pollution event. The health data are controlled for factors such as season and temperature referred to above, which can have major impacts on health, and when all such controls have been applied, the resultant day-to-day changes in morbidity or mortality are related to the air pollution measurements. These studies have proved extremely powerful in understanding the influence of air pollutants, especially particulate matter.

Nowadays most urban air pollutants have a common source in road traffic and since the main controls on concentrations are the rate of emission and the prevailing weather, traffic-generated pollutants tend to vary in much the same way from day-to-day. Therefore, a day with a high concentration of carbon monoxide will probably also have a high concentration of nitrogen dioxide and particulate matter. Consequently, disentangling the influences of the different pollutants when they vary from day-to-day in such a similar manner can be extremely difficult. Not all pollutants co-vary in this manner. For example, ozone tends to correlate positively with nitrogen dioxide in the summer but inversely in the winter months, hence making separation of the effects of the two pollutants much more straightforward when full annual data are utilized. The alternative cross-sectional cohort study design, as was used in the Harvard Six Cities study discussed later, relates measurements of air pollutant concentrations in different cities to the rates of mortality or morbidity in those cities. The Six Cities study was conducted using a time period over which traffic-generated pollutants were not as dominant as currently and therefore some separation of the effects of different pollutants was more straightforward.

Studies of occupationally exposed workers are particularly valuable in the case of chemical carcinogens, where frequently they provide the only source of real world data. In general, the effects of airborne chemical carcinogens are insufficiently large to be demonstrated through epidemioloical studies of the general population, and ethical considerations obviously rule out the use of chamber studies for chemical carcinogens. The residual problem with occupational studies is that the concentrations of air pollutants encountered far exceed those to which the general population are exposed, and extrapolation to lower concentrations is very much an act of faith. However, for genotoxic carcinogens there are reasonable grounds for believing that the dose–response function is approximately linear without any threshold and an assumption of this kind is

66

	Category	Pollutant	Threshold
Table 7 Categories of air pollutant toxic action	Irritants	Ozone, nitrogen dioxide, sulfur dioxide	May show a threshold
	Asphyxiant	Carbon monoxide	May show a threshold
	Genotoxic carcinogens	Benzene, 1,3-butadiene	Not believed to have a threshold
	Enzyme inhibitors	Lead	May show a threshold
	Mechanism uncertain	Particles	No threshold demonstrable at population level

generally inherent in the standard setting process, although for bodies such as the US EPA, who have taken quantitative risk assessment to a far greater extent in the standard setting process than other organizations, the use of other more sophisticated models is routine.

Finally, data from *animal studies* can occasionally prove useful. Such information usually has a value in demonstrating mechanisms rather than illuminating exposure–response functions. It may possibly throw light on the relative toxicity of chemical carcinogens for which human data are very sparse, but otherwise animal data have little direct use in the standard setting process.

5 Determining an Acceptable Level of Effects

This can often prove more difficult than defining exposure–response relationships. Table 7 attempts to categorize common air pollutants according to their mechanism of action. For each mechanism of action, different considerations come into play in looking for an acceptable level of effect.

Taking first the so-called 'irritant' pollutants, the approach taken by the WHO and EPAQS has been to determine a lowest observable effect level, taking into account sensitive groups where data are available, incorporation of a margin of safety to allow for more sensitive subjects than can take part in chamber studies and to set a standard which should be 'safe' for all groups. Since it is assumed that at least at an individual level the irritant pollutants have a threshold, it should be possible to set a standard protective of the kinds of individuals who took part in chamber studies, as well as some more sensitive groups through the incorporation of the margin of safety. An example of this approach is sulfur dioxide, where the key effect is that of bronchoconstriction in asthmatics. The lowest observable effect level from chamber studies is approximately 200 ppb over a few minutes. EPAQS[8] recommended to a standard of 100 ppb averaged over 15 minutes to incorporate a margin of safety and to allow for concentrations above this level occurring for periods of less than 15 minutes.

A second example is that of ozone, where the key effect is a reduction in lung capacity. EPAQS[5] considered the lowest observable effect level to be 80 ppb exposure over 6.6 hours, and incorporation of a margin of safety led to a standard of 50 ppb over eight hours. This approach assumes that there is a threshold for injury by ozone, although some have argued that this may not be the case, and indeed, some epidemiological data published after EPAQS made its recommendation suggest there is not. The concept of there being no threshold for adverse effects of ozone upon humans raises a very interesting paradox. Ozone concentrations are generally higher in rural than urban areas, and current daytime rural concentrations in the northern hemisphere typically show hourly averages of up to 50 ppb and eight hourly averages of 30–40 ppb in the absence of any severe pollution event.[14] This background of ozone arises from two sources. Around half of it comes from downward mixing of stratospheric ozone and is therefore wholly natural. The other half is the result of perturbation of the lower atmosphere by nitrogen oxides from anthropogenic combustion processes, which interact with methane and carbon monoxide to form ozone in sunshine. There is good reason to think that the human species and its antecedents in the evolutionary chain evolved in the presence of about 20 ppb of ozone, as this is a natural background which cannot be reduced without an undesirable reduction in stratospheric ozone concentrations. It seems unlikely that such a level of ozone would cause an adverse affect, and it is possible that the epidemiological studies suggesting that there is no threshold may be either insufficiently sensitive to recognize a threshold at this level, or may be seeing an effect which is the result of confounding by some other pollutant which varies in concentration in the same way as ozone. The latter is a distinct possibility, given the complexity of atmospheric photochemistry. Reduction of ground-level ozone concentrations to their pre-industrial level would require reductions in emissions of NO_x and VOC which would be so great as to change completely the nature of society as we know it. It seems most unrealistic to imagine that this would happen, and therefore if one accepts that there is either no threshold for the adverse effects of ozone or that the threshold is at the natural ozone background of around 20 ppb, then one must also accept that any realistically attainable air quality standard will involve health consequences from ozone for a small proportion of the population.

Carbon monoxide can be described as an asphyxiant since it takes up the oxygen carrying capacity of blood. In this case the key effect in standard setting has been the induction of angina in cardiovascular disease patients during exercise, for which the lowest observable effect level occurs at 3–4% carboxyhaemoglobin in blood. Incorporation of a safety margin aimed to ensure that carboxyhaemoglobin concentrations do not exceed 2.5% led EPAQS[6] to an air quality standard of 10 ppb over 8 hours exposure, and the WHO to a similar value for 8 hours and higher concentrations for shorter time periods.

For genotoxic carcinogens such as benzene, 1,3-butadiene and polycyclic aromatic hydrocarbons (PAHs), there is no totally safe level of exposure. The WHO have analysed the occupational disease studies and come up with an excess risk of contracting cancer following lifetime exposure. Taking for example the unit risk factor for benzene of $6 \times 10^{-6} (\mu g\,m^{-3})^{-1}$, this implies that six persons in a population of one million will contract cancer when exposed for their lifetime to a benzene concentration of $1\,\mu g\,m^{-3}$. Faced with the problem of setting a numerical air quality standard rather than an exposure–response gradient, EPAQS[4] used essentially the same occupational data to identify an exposure concentration below which a large cohort of workers showed no significant

[14] R.M. Harrison, in *Pollution: Causes, Effects & Control*, ed. R.M. Harrison, Royal Society of Chemistry, Cambridge, 1996.

excess of disease. In the case of benzene, this concentration was 500 ppb. EPAQS then divided by an exposure duration factor, which allows for the greater duration of exposure of the general population (*i.e.* 24 hours a day, 365 days a year) relative to the occupationally exposed population (40 hours per week, 46 weeks per year). The exposure duration factor used was 10. EPAQS divided also by a safety factor to protect sensitive groups, and adopted a factor of 10 for this. The application of both factors leads to an air quality standard of 5 ppb, and additionally a long-term target of 1 ppb was recommended. Both concentrations were expressed as running annual averages, reflecting the fact that it is long-term integrated exposure to chemical carcinogens rather than short-term excursions which are believed important in determining the induction of cancer. This methodology has been explained in depth elsewhere.[15,16]

The US EPA has done much to develop quantitative risk assessment of carcinogens for regulatory applications. Thus, quantitative estimates of carcinogenic potential akin to the unit risk factor cited above for benzene (which originates from the WHO) are used in standard setting. The attraction of this approach is that if unit risk factors can be established with confidence for a range of pollutants, and society can agree a tolerable level of risk, then maximum tolerable exposures to pollutants, and hence environmental quality standards, follow very straightforwardly. Such an approach has formed the basis for setting standards for many years in the field of radiological protection, where data collected from nuclear bomb survivors in Hiroshima and Nagasaki give a reasonable basis for estimation of unit risk factors. In the field of chemical carcinogenesis, the databases from which to estimate unit risk factors are far less substantial, and hence far greater uncertainty attaches to the estimates. Additionally, the unit risk factors obtained can be very sensitive to the model used to derive them, and hence in the UK, the Department of Health Committee on Carcinogenicity recommends against the routine use of quantitative risk assessment models. Hrudey and Krewski[17] have analysed some of the weaknesses in the modelling approaches used by the US EPA.

The next crucial question which has to be addressed is what is a tolerable level of risk? Naturally, there is no universally agreed answer to this question, and indeed, society is notably more tolerant of self-imposed risks (*e.g.* cigarette smoking) than of risks which are perceived as externally imposed such as outdoor air pollution. To put the matter into context, some representative levels of risk associated with well known events appear in Table 8. Given that it is not possible to achieve a situation of zero risk, long-term policy is generally directed at reducing risks to what is termed as a *de minimus* level, usually taken as a lifetime risk of one in 10^6. For short-term regulatory purposes a lifetime risk of one in 10^5 or one in 10^4 is seen as more realistic. Note that the risk estimates in Table 8 apply to annual risk and should therefore be adjusted by a factor of about 80 to give levels of lifetime risk. Thus, an annual risk of one in 10^6 equates to a lifetime risk of approximately one in 10^4. What does this mean in practice? A very approximate

[15] R. L. Maynard, K. M. Cameron, R. Fielder, A. McDonald and A. Wadge, *Hum. Exp. Toxicol.*, 1995, **14**, 175.
[16] R. J. Fielder, *Toxicology*, 1996, **113**, 222.
[17] S. E. Hrudey and D. Krewski, *Environ. Sci. Technol.*, 1995, **29**, 370A.

Table 8 Relative levels of annual risk

Descriptor	Risk estimate	Example	
High	$>1:100$	Transmission to susceptible household contact of measles and chickenpox	$1:1$–$1:2$
		Gastrointestinal effects of antibiotics	$1:10$–$1:20$
Moderate	$1:100$ to $1:1000$	Death from smoking 10 cigs/day	$1:200$
		Death, all natural causes, age 40 yrs	$1:850$
Low	$1:10^3$ to $1:10^4$	Death from influenza	$1:5000$
		Death in road accidents	$1:8000$
Very low	$1:10^4$ to $1:10^5$	Death from leukaemia	$1:12\,000$
		Death, playing soccer	$1:25\,000$
		Death, accident at home	$1:26\,000$
		Death, accident at work	$1:43\,000$
		Death from homicide	$1:100\,000$
Minimal	$1:10^5$ to $1:10^6$	Death, accident on railway	$1:500\,000$
Negligible	$<1:10^6$	Death, hit by lightning	$1:10^7$
		Death, release of radiation from nuclear power station	$1:10^7$

Based on ref. 18.

ilustration can be taken from the EPAQS benzene standard, which was set up at 5 ppb ($16\,\mu g\,m^{-3}$); when combined with the WHO unit risk factor this implies a lifetime risk of 1×10^{-4}, which is comparable with the levels of risk considered tolerable in radiological protection. However, this numerical exercise can give a highly spurious impression of precision to the process and it would be quite wrong for quantitative risk assessment to be regarded as a universal panacea in standard setting.

Perhaps the most difficult of all non-threshold pollutants is particulate matter. The literature contains a substantial body of epidemiological studies, linking a range of adverse health outcomes with day-to-day changes in the concentration of particulate matter within a city, generally measured as PM_{10}. These so-called time series studies give no indication of a threshold concentration below which no effects occur, and indeed, Watt and co-workers[19] have argued that although there will be a threshold concentration for individuals below which no harm from particle exposure will occur, this threshold will vary considerably between individuals. Additionally, within a given city, true individual exposure to particulate matter will also vary substantially. Therefore, within the population of a city, because of the wide distribution of individual thresholds and individual exposures, no threshold will be observable. In addition to the time series studies, there are three important cross-sectional studies which have looked at the rates of

[18] Department of Health, *Chief Medical Officer's Report for 1995*, HMSO, London, 1996.
[19] M. Watt, D. Golden, J. Cherrie and A. Seaton, *Occup. Environ. Med.*, 1995, **52**, 790.

mortality and disease in populations with different long-term exposures to particulate matter, and have shown appreciably elevated death rates in the cities with high fine particle concentrations. The interpretation of these results in relation to loss of life expectancy is extremely difficult, and furthermore, whilst the time series studies show that more people die and are admitted to hospital for respiratory and cardiovascular diseases on high particulate matter pollution days, it is unclear whether these events are simply being advanced by a few days, months or years. Hence the impact of particulate matter pollution on the prevalence of disease and reduction of life expectancy in the population as a whole is unclear. Reflecting these uncertainties, the WHO[11] produced a series of tables expressing exposure–response functions for a range of outcomes such as bronchodilator use, cough, lower respiratory symptoms, respiratory hospital admissions and mortality. Faced with the problem of setting a numerical standard, EPAQS[9] acknowledged that it was impossible to set a standard which would be totally protective against all adverse effects and recommended a standard for PM_{10} of $50 \,\mu g \, m^{-3}$, 24 hour running mean, as a concentration at which health effects on individuals were likely to be small and the very large majority of individuals will be unaffected. It was noted that a rise from a daily average PM_{10} level of $20 \,\mu g \, m^{-3}$ to $50 \,\mu g \, m^{-3}$, a concentration which was exceeded on average one day in 10 in a study in Birmingham, UK, would be expected to be associated with just over one extra patient on average being admitted to hospital with respiratory disease daily in a population of one million. This was considered tolerable, although clearly this is a very subjective judgement.

The standard setting process described above for particulate matter is based solely on the results of the many published time series studies of air pollution and daily morbidity and mortality. There have also been published three long-term studies[20-22] which give some indication of the effects of chronic exposure to particulate matter over many years. The studies are all cohort studies which take a group of known individuals and follow them forwards in time, accumulating data upon the morbidity and mortality amongst the cohort, who are assessed initially in relation to individual risk factors such as smoking, body mass index and socio-economic status. The studies are cross-sectional in the sense that the subjects studied live in a number of cities (from six in the Harvard Six Cities Study[20] to 151 in the so-called American Cancer Society Study[21]) with differing levels of air pollution. Two of the studies have revealed linear relationships between mortality rates and airborne concentrations of particulate matter measured as PM_{10} or $PM_{2.5}$ after normalizing the data for the other risk factors. The outcomes of these two studies[20,21] suggest major differences in life expectancy due to ambient particle exposure, but without making a number of arbitrary assumptions it is not possible to estimate the number of years of life lost. A further difficulty in using the studies to set air quality standards for particulate

[20] D. W. Dockery, C. A. Pope III, X. Xu, J. D. Spengler, H. J. Ware, M. E. Fay, B. G. Ferris Jr. and F. E. Speizer, *New Engl. J. Med.*, 1993, **329**, 1753.

[21] C. A. Pope III, M. J. Thun, M. W. Namboodiri, W. D. Dockery, J. S. Evans, F. E. Speizer and C. W. Heath, *Am. J. Respir. Crit. Care Med.*, 1995, **151**, 669.

[22] D. E. Abbey, M. D. Lebowitz, P. K. Mills, F. F. Petersen, W. L. Beeson and R. J. Burchette, *Inhal. Toxicol.*, 1995, **7**, 18.

matter is that it is probable that the subjects' exposures having the greatest long-term health impact will have taken place before the commencement of the study and before monitoring data became available. It is in this earlier time frame that concentrations will have been at their highest, and the subjects of the study in their infancy when long-term detriments to their well-being from air pollutant exposure would be greatest. Therefore, despite apparently revealing quite major impacts of particulate matter on health, these studies give little basis for determining exposure–response relationships or for standard setting. A further difficulty, as with the time series studies, is that the cohort studies to date show no threshold below which effects on health are not observable.

6 Concluding Remarks

All air quality standards should combine both a concentration and an averaging time. That averaging time reflects the duration of exposure associated with the eliciting of a response from exposure to the pollutant. Thus, for the irritant gases the exposure times are generally relatively short as the effects are acute, whilst for the genotoxic carcinogens the effects are chronic and the averaging times are long. The question also arises as to what locations the standards should be applied. The UK position on this is both rational and perceptive in that the UK National Air Quality Strategy[12] states that 'the objectives should apply in non-occupational, near ground-level outdoor locations where a person might reasonably be expected to be exposed over the relevant averaging period'. This is very important in implicitly including hotspot locations where people spend a significant amount of time in relation to the pollutant and potential effects, but excluding extreme situations that have no real relevance to human health. Clearly, monitoring strategies should be designed to reflect this kind of logic.

Finally, it must be recognized that pollutants can have interactive effects and may possibly act synergistically. Thus, for example, at one time it was believed that smoke and sulfur dioxide acted synergistically to elicit a greater effect than the sum of the two acting independently. This was largely a matter of faith as at the time the two pollutants had a major common source in coal combustion, and when one pollutant was elevated in concentration, the other was also. More recent thinking has suggested that the two pollutants tend to act independently of one another. A recent authoritative report[23] reviewing literature evidence for the health impact of pollutant mixtures concluded that there was little hard evidence available to suggest that pollutant interactions were particularly important. However, recent epidemiological studies have failed to disentangle in a wholly consistent manner the impacts of the many different pollutants whose concentrations tend to co-vary in the atmosphere due to common sources (usually motor traffic) and the same meteorological influences on concentation. Thus, although some studies appear to point to individual pollutants, others are more equivocal, and some studies have been totally unable to disaggregate clearly the effects of the various pollutants in the urban mix. A view is therefore commonly being expressed that the health effects being identified result from exposure to the mix

---

[23] Department of Health, Advisory Group on the Medical Aspects of Air Pollution Episodes, *Health Effects of Exposures to Mixtures of Air Pollutants*, HMSO, London, 1995.

as a whole and that attribution to individual components should be conducted with considerable caution. Thus, there are a variety of expert views on this matter, which is itself a very important one in standard setting. It is to be hoped that this issue will clarify over the next few years as epidemiological studies improve in their sophistication.

The 1997 US EPA Standards for Particulate Matter and Ozone

MORTON LIPPMANN

1   Introduction

Under the mandate of the Clean Air Act (CAA) of the United States, as amended in 1977, the Administrator of the Environmental Protection Agency (EPA) is supposed to review the basis for its National Ambient Air Quality Standards (NAAQS) every five years and, if necessary for the protection of public health and/or welfare, issue new or revised standards. In the past, as shown in Table 1, the ozone (O_3) and particulate matter (PM) NAAQS have been revised, albeit not at the specified five-year intervals. The 1997 revisions to the NAAQS for PM and O_3 were unusual in NAAQS revisions in a number of respects, including especially: (1) their simultaneous promulgations; (2) the historically tight timetables involved in their preparation and public review; (3) the extraordinary controversy they engendered during the review process and since their promulgation; and (4) the substantial increases in the number of additional communities that will not be in compliance with the NAAQS for both PM and O_3.

The CAA requires the EPA Administrator to promulgate NAAQS that protect the health of sensitive segments of the public with an adequate 'margin of safety'. In addition to the health-based (primary) standards, there are also welfare-based (secondary) standards, referring to human comfort, economic damage to materials, crops, livestock, and ecological balance. The CAA is unique among the statutes that EPA enforces in that the primary NAAQS are supposed to be set without regard to the costs they impose on society. However, EPA does perform cost–benefit analyses for NAAQS to conform with a Presidential directive requiring such analyses for all regulations that significantly affect the national economy.

In the process of considering the revision of a NAAQS, the National Center for Environmental Assessment (NCEA) of the EPA's Office of Research and Development (ORD) prepares a Criteria Document (CD). This is a massive and encyclopedic summary of the peer-reviewed literature that bears upon the chemical and physical nature of the pollutant, its measurement in the ambient air, its primary sources and secondary transformations in the atmosphere, its

Table 1 1997 Revisions: US National Ambient Air Quality Standards (NAAQS)

I. Ozone (Revision of NAAQS Set in 1979 and Reaffirmed in 1993)

	1979 NAAQS	1997 NAAQS
Daily concentration limit (ppb)	120	80
Averaging time	maximum: 1 h av.	maximum: 8 h av.
Basis for excessive concentration	4th highest over 3 year period	3 year av. of 4th highest in each year
Equivalent stringency for 1 h max in new format (ppb)	~ 90	
Number of US counties expected to exceed NAAQS	106	280
Number of people in counties exceeding NAAQS	74×10^6	113×10^6

II. Particulate Matter (Revision of NAAQS Set in 1987)

	1987 NAAQS	1997 NAAQS	
Index pollutant	PM_{10}	PM_{10}	$PM_{2.5}$
Annual av. concentration limit ($\mu g\, m^{-3}$)	50	50	15
Daily concentration limit ($\mu g\, m^{-3}$)	150	150	65
Basis for excessive daily concentration	4th highest over 3 year period	>99th percentile av. over 3 years	>98th percentile av. over 3 years
Number of US counties expected to exceed NAAQS	41	14	~ 150
Number of people in counties exceeding NAAQS	29×10^6	$\sim 9 \times 10^6$	$\sim 68 \times 10^6$

transport and fate, its geographic distribution, methods of sampling and analysis, the exposures of populations and other receptors to the pollutant (both outdoors and indoors), its dosimetry once inhaled or deposited, the effects that it produces, and, in the most recent CDs, an integrated summary of sources, exposures, and responses.

EPA's plans for the organization of the CD are reviewed by the Clean Air Scientific Advisory Committee (CASAC), an independent external advisory committee created in response to a mandate of the CAA amendments of 1977.

Subsequently, external review drafts of the CD are reviewed in public by CASAC, which then makes recommendation for changes in the draft. There is typically a second external review draft and further CASAC and public commentary. When CASAC is satisfied that the CD is a complete and balanced summary and analysis, it prepares a 'closure letter' to the Administrator endorsing the document as a suitable foundation for setting a NAAQS. CASAC also publicly reviews a companion document prepared by EPA's Office of Air Quality Planning and Standards (OAQPS). This document makes selective use of the contents of the CD that are most relevant to the selection of critical elements in the NAAQS, such as the indicator pollutant(s), averaging times, method(s) for analysis of concentration, sensitive segments of the population, exposure–response relationships, and margin-of-safety associated with several alternative NAAQS levels under consideration. The final draft of this companion document, known as a Staff Paper (SP), is prepared after CASAC Closure on the CD. After a final public review, the CASAC chair prepares a closure letter on the SP.

The preparation of an economic impact analysis and further internal reviews in EPA and the federal Office of Management and Budget (OMB) precedes a Federal Register proposal to change or reaffirm an existing NAAQS. After a further public comment period, the EPA Administrator announces the promulgation of the NAAQS in the Federal Register.

This exhaustive review process by EPA Staff (ECAO and OAQPS) and CASAC inevitably identifies important knowledge gaps that limit confidence in the NAAQS options as optimal choices. Recognizing these limitations, EPA Staff prepares lists of research recommendations for public review by CASAC, and CASAC summarizes and prioritizes research needs in a letter report to the EPA Administrator. It is hoped that EPA and other research sponsors will then arrange to support research that can close the critical knowledge gaps, so that the next cycle of NAAQS reviews can lead to more refined and well targeted NAAQS.

The author has served on all of the CASAC panels reviewing CDs, SPs, and research needs for the PM and O_3 NAAQS since 1980 in various capacities (consultant 1980–1982; member 1982–1987; chairman 1983–1987; consultant 1988–1997). The views expressed herein are those of the author, and do not necessarily reflect the views of the EPA.

2 Particulate Matter

In Europe, and elsewhere in the eastern hemisphere, particulate pollution has generally been measured as black smoke (BS) in terms of the optical density of stain caused by particles collected on a filter disc. However, it has been expressed in gravimetric terms ($\mu g\,m^{-3}$) based on standardized calibration factors. By contrast, US standards have specified direct gravimetric analyses of filter samples collected by a reference sampler built to match specific physical dimensions or performance criteria.

While justifications for the specific measurement techniques that have been used have generally been based on demonstrated significant quantitative associations between the measured quantity and human mortality, morbidity, or lung function differences, it is fair to say that we still lack established biological

Figure 1 Representative example of a mass distribution of ambient PM as function of aerodynamic particle diameter. A wide ranging aerosol classifier (WRAC) provides an estimate of the full coarse mode distribution. Inlet restrictions of the TSP high volume sampler, the PM_{10} sampler, and the $PM_{2.5}$ sampler reduce the integral mass reaching the sampling filter (Adapted from ref. 5)

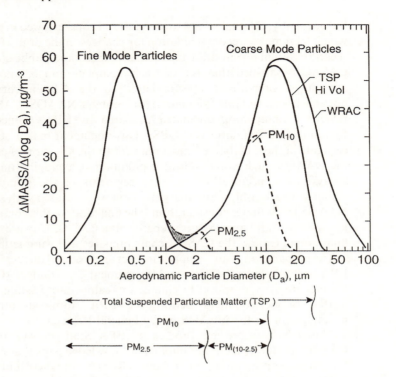

mechanisms that could account for these associations, and that we have too little information on the relative toxicities of the myriad specific constituents of airborne PM. In addition to chemical composition, airborne PM also varies in particle size distribution, which affects the number of particles that reach target sites as well as the particle surface area. To date, there are no NAAQS for PM constituents (other than lead) or for number or surface concentrations.

A broad variety of processes produce suspended particulate matter (PM) in the ambient air in which we live and breathe, and there is an extensive body of epidemiological literature that demonstrates that there are statistically significant associations between the concentrations of airborne PM and the rates of mortality and morbidity in human populations. The PM concentrations have almost always been expressed in terms of mass, although one recent study indicates that number concentration may correlate better with effects than does fine particle mass.[1] In those studies that reported on associations between health effects and more than one mass concentration, the strength of the association generally improves as one goes from total suspended particulate matter (TSP) to thoracic particulate matter, *i.e.*, PM less than 10 μm in aerodynamic diameter (PM_{10}), to fine particulate matter, *i.e.*, PM less than 2.5 μm in aerodynamic diameter ($PM_{2.5}$). The influence of a sampling system inlet on the sample mass collected is illustrated in Figure 1.

[1] A. Peters, E. Wichmann, T. Tuch, J. Heinrich, and J. Heyder, *Am. J. Respir. Crit. Care Med.*, 1997, in press.

The PM$_{2.5}$ distinction, while nominally based on particle size, is in reality a means of measuring the total gravimetric concentration of several specific chemically distinctive classes of particles that are emitted into or formed within the ambient air as very small particles. In the former category (emitted) are carbonaceous particles in wood smoke and diesel engine exhaust. In the latter category (formed) are carbonaceous particles formed during the photochemical reaction sequence that also leads to ozone formation, as well as the sulfur and nitrogen oxide particles resulting from the oxidation of sulfur dioxide and nitrogen oxide vapors released during fuel combustion and their reaction products.

The coarse particle fraction, *i.e.*, those particles with aerodynamic diameters larger than $\sim 2.5\,\mu$m, are largely composed of soil and mineral ash that are mechanically dispersed into the air. Both the fine and coarse fractions are complex mixtures in a chemical sense. To the extent that they are in equilibrium in the ambient air, it is a dynamic equilibrium in which they enter the air at about the same rate as they are removed. In dry weather, the concentrations of coarse particles are balanced between dispersion into the air, mixing with air masses, and gravitational fallout, while the concentrations of fine particles are determined by rates of formation, rates of chemical transformation, and meteorological factors. PM concentrations of both fine and coarse PM are effectively depleted by rainout and washout associated with rain. Further elaboration of these distinctions is provided in Table 2.

In the absence of any detailed understanding of the specific chemical components responsible for the health effects associated with exposures to ambient PM, and in the presence of a large and consistent body of epidemiological evidence associating ambient air PM with mortality and morbidity that cannot be explained by potential confounders such as other pollutants, aeroallergens, or ambient temperature or humidity, the EPA has established standards based on mass concentrations within certain prescribed size fractions (see Table 1).

As indicated in Table 2, fine and coarse particles generally have distinct sources and formation mechanisms, although there may be some overlap. Primary fine particles are formed from condensation of high temperature vapors during combustion. Secondary fine particles are usually formed from gases in three ways: (1) nucleation (*i.e.*, gas molecules coming together to form a new particle); (2) condensation of gases onto existing particles; and (3) by reaction of absorbed gases in liquid droplets. Particles formed from nucleation also coagulate to form relatively larger aggregate particles or droplets with diameters between 0.1 and 1.0 μm, and such particles normally do not grow into the coarse mode. Particles form as a result of chemical reaction of gases in the atmosphere that lead to products that either have a low enough vapor pressure to form a particle, or react further to form a low vapor pressure substance. Some examples include: (1) the conversion of sulfur dioxide (SO$_2$) to sulfuric acid droplets (H$_2$SO$_4$); (2) reactions of H$_2$SO$_4$ with ammonia (NH$_3$) to form ammonium hydrogensulfate (NH$_4$HSO$_4$) and ammonium sulfate [(NH$_4$)$_2$SO$_4$]; (3) the conversion of nitrogen dioxide (NO$_2$) to nitric acid vapor (HNO$_3$), which reacts further with NH$_3$ to form particulate ammonium nitrate (NH$_4$NO$_3$). Although some directly emitted particles are found in the fine fraction, particles formed secondarily from gases dominate the fine fraction mass.

Table 2 Comparisons of ambient fine and coarse mode particles

	Fine mode	Coarse mode
Formed from	Gases	Large solids/droplets
Formed by	Chemical reaction; nucleation; condensation; coagulation; evaporation of fog and cloud droplets in which gases have dissolved and reacted	Mechanical disruption (*e.g.*, crushing, grinding, abrasion of surfaces); evaporation of sprays; suspension of dusts
Composed of	Sulfate, SO_4^{2-}; nitrate, NO_3^-; ammonium, NH_4^+; hydrogen ion, H^+; elemental carbon; organic compounds (*e.g.*, PAHs, PNAs); metals (*e.g.*, Pb, Cd, V, Ni, Cu, Zn, Mn, Fe); particle-bound water	Resuspended dusts (*e.g.*, soil dust, street dust); coal and oil fly ash; metal oxides of crustal elements (Si, Al, Ti, Fe); $CaCO_3$, NaCl, sea salt; pollen, mold spores; plant/animal fragments; tire wear debris
Solubility	Largely soluble, hygroscopic, and deliquescent	Largely insoluble and non-hygroscopic
Sources	Combustion of coal, oil, gasoline, diesel, wood; atmospheric transformation products of NO_x, SO_2, and organic compounds including biogenic species (*e.g.*, terpenes); high temperature processes, smelters, steel mils, *etc.*	Resuspension of industrial dust and soil tracked onto roads; suspension from disturbed soil (*e.g.*, farming, mining, unpaved roads); biological sources; construction and demolition; coal and oil combustion; ocean spray
Lifetimes	Days to weeks	Minutes to hours
Travel distance	100s to 1000s of kilometers	<1 to 10s of kilometers

Source: EPA.

By contrast, most of the coarse fraction particles are emitted directly as particles, and result from mechanical disruption such as crushing, grinding, evaporation of sprays, or suspensions of dust from construction and agricultural operations. Basically, most coarse particles are formed by breaking up bigger masses into smaller ones. Energy considerations normally limit coarse particle sizes to greater than $1.0\,\mu m$ in diameter. Some combustion-generated mineral particles, such as fly ash, are also found in the coarse fraction. Biological material such as bacteria, pollen, and spores may also be found in the coarse mode. As a result of the fundamentally different chemical compositions and sources of fine and coarse fraction particles, the chemical composition of the sum of these two fractions, PM_{10}, is more heterogenous than either mode alone.

Figure 2 presents a synthesis of the available published data on the chemical composition of $PM_{2.5}$ and coarse fraction particles in US cities by region. Each

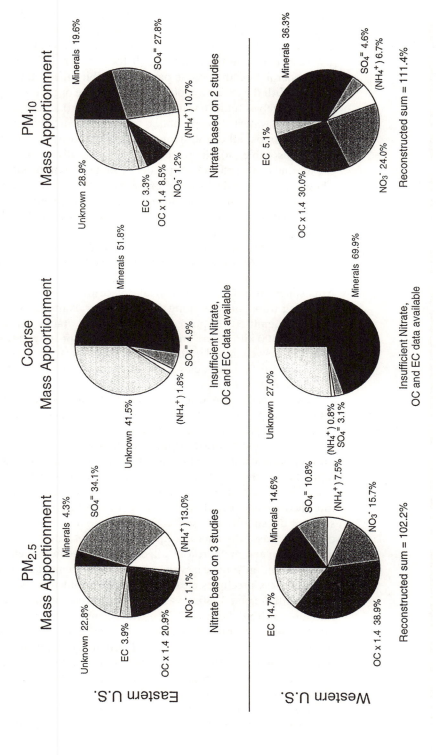

Figure 2 Representative distributions of major components of sampled PM in the fine fraction (PM$_{2.5}$), the coarse component of PM$_{10}$, and PM$_{10}$ as a whole, respectively, in the eastern US (top) and western US (bottom) (Adapted from ref. 5)

fraction also has regional patterns resulting from the differences in sources and atmospheric conditions. In addition to the larger relative shares of crustal materials in the west, total concentrations of coarse fraction particles are generally higher in the arid areas of the western and south-western US.

In general, fine and coarse particles exhibit different degrees of solubility and acidity. With the exception of carbon and some organic compounds, fine particle mass is largely soluble in water and hygroscopic (*i.e.*, fine particles readily take up and retain water). Except under fog conditions, the fine particle mode also contains almost all of the strong acid. By contrast, coarse mineral particles are mostly insoluble, non-hygroscopic, and generally basic.

Fine and coarse particles typically exhibit different behavior in the atmosphere. These differences affect several exposure considerations, including the representativeness of central-site monitored values and the behavior of particles that were formed outdoors after they penetrate into homes and buildings where people spend most of their time.

Fine accumulation mode particles typically have longer atmospheric lifetimes (*i.e.*, days to weeks) than coarse particles, and tend to be more uniformly dispersed across an urban area or large geographic region, especially in the eastern US. Atmospheric transformations can take place locally, during atmospheric stagnation, or during transport over long distances. For example, the formation of sulfates from SO_2 emitted by power plants with tall stacks can occur over distances exceeding 300 kilometers and 12 hours of transport time; therefore, the resulting particles are well mixed in the air shed. Once formed, the very low dry deposition velocities of fine particles contribute to their persistence and uniformity throughout an air mass.

Larger particles generally deposit more rapidly than small particles; as a result, total coarse particle mass will be less uniform in concentration across a region than are fine particles. The larger coarse particles ($> 10 \mu$m) tend to rapidly fall out of the air and have atmospheric lifetimes of only minutes to hours, depending on their size, wind velocity, and other factors. Their spatial impact is typically limited by a tendency to fall out in the nearby downwind area. The atmospheric behavior of the smaller particles within the 'coarse fraction' ($PM_{10-2.5}$) is intermediate between that of the larger coarse particles and fine particles. Thus, some of the smaller coarse fraction particles may have lifetimes on the order of days and travel distances of up to 100 km or more. In some locations, source distribution and meteorology affects the relative homogeneity of fine and coarse particles, and in some cases the greater measurement error in estimating coarse fraction mass precludes clear conclusions about relative homogeneity.

Up until the mid-1980s, available PM concentrations in the US were generally measured as TSP. Because TSP includes, and can be dominated by, particles too large to penetrate into the thorax, it is a poor index of inhalation hazard. Since the dispersion of large particles is limited, proximity of the sampler to local sources of dust has a major influence on measured TSP concentrations. The artifacts also vary with season and climate, and can be especially severe in the arid portions of the western US.

Despite the inherent limitations of (1) the assumption of equivalent toxicity of all sampled particles, and (2) the sampling and analytical artifacts that limit the accuracy and precision of measured PM concentrations, there is a substantial body of epidemiological evidence for statistically significant associations between airborne PM concentrations and excess mortality and morbidity. Furthermore, the mortality and morbidity effects appear to be coherent and not explicable on the basis of known potential confounding factors or co-existing gas phase pollutants.[2,3]

Review of the Health Effects Literature that Influenced the New PM NAAQS

During the 1990s, there was a great increase in the number of peer-reviewed papers describing time-series studies of the associations between daily ambient air pollutant concentrations and daily rates of mortality and hospital admissions for respiratory diseases. Also, results of two prospective cohort studies of annual mortality rates were published. In terms of morbidity, there has been a rapid growth of the literature showing associations between airborne particle concentrations and exacerbation of asthma, increased symptom rates, decreased respiratory function, and restricted activities.

Table 3 shows an analysis of acute mortality studies in nine communities with measured PM_{10} concentrations.[4] As indicated in this table, the coefficients of response tend to be higher when the PM_{10} is expressed as a multiple-day average concentration, and lower when other air pollutants are included in multiple-regression analyses. In any case, the results in each city (except for the very small city of Kingston, TN) indicate a statistically significant association. It is also clear from recent research that the associations between PM_{10} and daily mortality are not seriously confounded by weather variables or the presence of other criteria pollutants.[5] Figure 3 shows that the calculated relative acute mortality risks for PM_{10} are relatively insensitive to the concentrations of SO_2, NO_2, CO, and O_3. The results are also coherent as described by Bates.[2] Figure 4 shows that the relative risks (RRs) for respiratory mortality are greater than for total mortality and hospital admissions.

While there is mounting evidence that excess daily mortality is associated with short-term peaks in PM_{10} pollution, the public health implications of this evidence are not yet fully clear. Key questions remain, including:

- Which specific components of the fine particle fraction ($PM_{2.5}$ and coarse particle fraction of PM_{10} are most influential in producing the responses?
- Do the effects of the PM_{10} depend on co-exposure to irritant vapors, such as ozone, sulfur dioxide, or nitrogen oxides?
- What influences do multiple-day pollution episode exposures have on daily responses and response lags?

[2] D. V. Bates, *Environ. Res.*, 1992, **59**, 336.
[3] C. A. Pope, Jr., D. W. Dockery, and J. Schwartz, *Inhal. Toxicol.*, 1995, **7**, 1.
[4] G. D. Thurston, personal communication of table prepared for draft EPA Criteria Document on Particulate Matter, 1995.
[5] US EPA, *Air Quality Criteria for Particulate Matter*, EPA/600/P-95/001F, US Environmental Protection Agency, Washington, DC, 1996.

Table 3 Comparison of time-series study estimates of total mortality relative risk (RR) for a $100\,\mu g\,m^{-3}$ PM_{10} increase

Study area (reference)	Measured PM_{10} concentrations		RR for $100\,\mu g\,m^{-3}$	95% CI for $100\,\mu g\,m^{-3}$
	Mean ($\mu g\,m^{-3}$)	Maximum ($\mu g\,m^{-3}$)		
1. Utah Valley, UT (Pope *et al.*, 1992)	47	297	1.16[a,d]	(1.10–1.22)
2. St. Louis, MO (Dockery *et al.*, 1992)	28	97	1.16[a,c]	(1.01–1.33)
3. Kingston, TN (Dockery *et al.*, 1992)	30	67	1.17[a,c]	(0.88–1.57)
4. Birmingham, AL (Schwartz, 1993)	48	163	1.11[a,d]	(1.02–1.20)
5. Athens, Greece (Touloumi *et al.*, 1994)	78	306	1.07[a,c] 1.03[b,c]	(1.05–1.09) (1.00–1.06)
6. Toronto, Canada (Özkaynak *et al.*, 1994)	40	96	1.07[a,c] 1.05[b,c]	(1.05–1.09) (1.03–1.07)
7. Los Angeles, CA (Kinney *et al.*, 1995)	58	177	1.05[a,c] 1.04[b,c]	(1.00–1.11) (0.98–1.09)
8. Chicago, IL (Ito *et al.*, 1995)	38	128	1.05[b,c]	(1.01–1.10)
9. Santiago, Chile (Ostro *et al.*, 1995)	115	367	1.08[a,c] 1.15[a,d]	(1.06–1.12) (1.08–1.22)

[a]Single pollutant model (*i.e.* PM_{10}). [b]Multiple pollutant model (*i.e.* PM_{10} and other pollutants simultaneously). [c]One-day mean PM_{10} concentration employed. [d]Multiple-day mean PM_{10} concentration employed.
From M. Lippmann, in *Aerosol Inhalation: Recent Research Frontiers*, ed. J.C.M. Marijnissen and L. Gradon, Kluwer Academic Publishers, 1996, pp. 1–25.

- Does long-term chronic exposure predispose sensitive individuals being 'harvested' on peak pollution days?
- How much of the excess daily mortality is associated with life-shortening measured in days or weeks *vs.* months, years, or decades?

The last question above is a critical one in terms of the public health impact of excess daily mortality. If, in fact, the bulk of the excess daily mortality were due to 'harvesting' of terminally ill people who would have died within a few days, then the public health impact would be much less than if it led to prompt mortality among acutely ill persons who, if they did not die then, would have recovered and lived productive lives for years or decades longer. An indirect answer to this question is provided by the results of two relatively recent prospective cohort studies of annual mortality rates in relation to long-term pollutant exposures.

Dockery *et al.*[6] reported on a 14- to 16-year mortality follow-up of 8111 adults in six US cities in relation to average ambient air concentrations of TSP $PM_{2.5}$, fine particle SO_4^{2-}, O_3, SO_2, and NO_2. Concentration data for most of these pollutant variables were available for 14–16 years. The mortality rates were adjusted for cigarette smoking, education, body mass index, and other influential

[6] D.W. Dockery, C.A. Pope III, X. Xu, J.D. Spengler, J.H. Ware, M.E. Fay, B.G. Ferris, Jr., and F.E. Speizer, *New Engl. J. Med.*, 1993, **329**, 1753.

Figure 3 Relationship between RR associated with PM_{10} and peak daily levels of other criteria pollutants (Adapted from ref. 5)

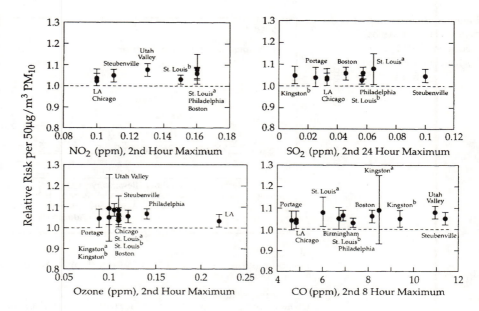

factors not associated with pollution. The two pollutant variables that best correlated with total mortality (which was mostly attributable to cardiopulmonary mortality) were $PM_{2.5}$ and SO_4^{2-}. The overall mortality rate ratios were expressed in terms of the range of air pollutant concentrations in the six cities. The rate-ratios (and 95% confidence intervals) for both $PM_{2.5}$ and SO_4^{2-} were 1.26 (1.08–1.47) overall and 1.37 (1.11–1.68) for cardiopulmonary. The mean life-shortening was in the range of 2–3 years.

Pope *et al.*[7] linked SO_4^{2-} data from 151 US metropolitan areas in 1980 with individual risk factors on 552 138 adults who resided in these areas when enrolled in a prospective study in 1982, as well as $PM_{2.5}$ data for 295 223 adults in 50 communities. Deaths were ascertained through December 1989. The relationships of air pollution to all-cause, lung cancer, and cardiopulmonary mortality were examined using multivariate analysis which controlled for smoking, education, and other risk factors. Particulate air pollution was associated with cardiopulmonary and lung cancer mortality, but not with mortality due to other causes. Adjusted relative risk ratios (and 95% confidence intervals) of all-cause mortality for the most polluted areas compared with the least polluted equaled 1.15 (1.09–1.22) and 1.17 (1.09–1.26) when using SO_4^{2-} and $PM_{2.5}$, respectively. The mean life-shortening in this study was between 1.5 and 2 years. The results were similar to those found in the previous cross-sectional studies of Özkaynak and Thurston[8] and Lave and Seskin.[9] Thus, the results of these earlier studies provide some confirmatory support for the findings of Pope *et al.*,[7] whose results indicate that the concerns about the credibility of the earlier results, due to their inability

[7] C. A. Pope, III, M. J. Thun, M. Namboodiri, D. W. Dockery, J. S. Evans, F. E. Speizer, and C. W. Heath, Jr., *Am. J. Respir. Crit. Care Med.*, 1995, **151**, 669.
[8] H. Özkaynak and G. D. Thurston, *Risk Anal.*, 1987, **7**, 449.
[9] L. B. Lave and E. P. Seskin, *Science*, 1970, **169**, 723.

Figure 4 Relationships between relative risks per $50\,\mu g\,m^{-3}$ PM$_{10}$ and health effects (Adapted from ref. 5)

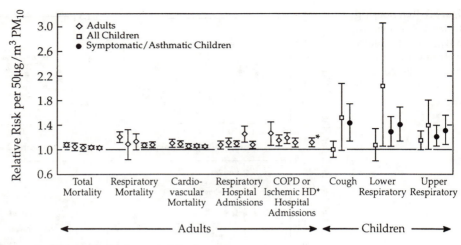

Total, Respiratory, Cardiovascular Mortality
    1. Pope et al. (1992)   2. Schwartz (1993)   3. Styer et al. (1995)   4. Ostro et al. (1995a)
    5. Ito and Thurston (1996)

Respiratory Hospital Admissions
    1. Schwartz (1995) New Haven, CT   2. Schwartz (1995) Tacoma,WA   3. Schwartz (1996) Spokane, WA
    4. Ito and Thurston (1994) Toronto, Canada

COPD or Ischemic HD* Hospital Admissions
    1. Schwartz (1994f) Minneapolis, MN   2. Schwartz (1994c) Birmingham, AL
    3. Schwartz (1996) Spokane, WA   4. Schwartz (1994d), Detroit, MI
    *5. Schwartz & Morris (1995), Detroit, MI, Ischemic HD

Cough, Lower Respiratory, Upper Respiratory
    1. Hoek and Brunekreef (1993)   2. Styer et al. (1994)   3. Pope & Dockery (1992), symptomatic children

to control for potentially confounding personal factors such as smoking and socioeconomic variables, can be eased.

The Dockery et al.[6] study had the added strength of data on multiple PM metrics. As shown in Figure 5, the association becomes stronger as the PM metric shifts from TSP to PM$_{10}$. Within the thoracic fraction (PM$_{10}$), the association is much stronger to the fine particle component (PM$_{2.5}$) than for the coarse component. Within the PM$_{2.5}$ fraction, both the SO$_4^{2-}$ and non-SO$_4^{2-}$ fractions correlate very strongly with annual mortality, suggesting a non-specific response to fine particles.

If, in fact, more people are dying of cardiopulmonary causes on a given day because of exposures to elevated concentrations of PM, it would be reasonable to expect higher daily rates of emergency hospital admissions and visits to emergency rooms and clinics for similar causes. This expectation is consistent with the results summarized in Figure 4. These studies indicate that indices of PM, such as daily concentrations of PM$_{10}$, SO$_4^{2-}$, and BS, are generally significantly associated with excess daily emergency admissions to hospitals for either respiratory diseases or cardiac diseases, or both. These studies have not shown associations with non-cardiopulmonary causes, and the influence of PM has generally been found to remain in multiple regression analyses that included other criteria pollutants. However, for respiratory diseases, the influence of

Figure 5 Adjusted relative risks for annual mortality are plotted against each of seven long-term average particle indices in the Six City Study, from largest size range [total suspended particulate matter (lower left), through sulfate and non-sulfate fine particle concentrations (upper right)]. Note that a relatively strong linear relationship is seen for fine particles, and for its sulfate and non-sulfate components. Topeka (T), which has a substantial coarse particle component of thoracic particle mass, stands apart from the linear relationship between relative risk and thoracic particle concentration (Adapted from ref. 5)

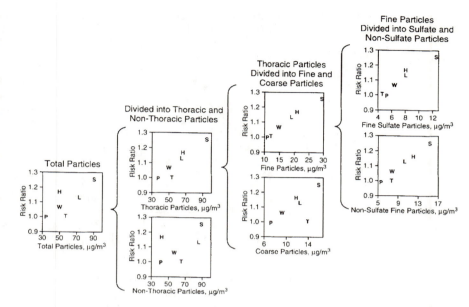

summertime O_3 has generally been greater than that of PM. This is in contradistinction to excess daily mortality, where the influence of PM is generally much greater than that of O_3. For hospitalizations for cardiac diseases, the most influential criteria pollutants appear to be PM and CO.

The importance of the fine particles as a risk factor for subnormal vital capacity in children is illustrated in Figure 6, which shows data collected in the Harvard–Health Canada cross-sectional study of 22 US and Canadian communities.[10] There was a significant association between the percentage of children with forced vital capacity (FVC) < 85% of predicted and fine particle mass concentration, but no apparent association with the coarse component of PM_{10}. Actually, the strongest association observed in this comparison was for the H^+ component of the fine particles. Most of the recent epidemiological studies have not had the advantage of available $PM_{2.5}$, SO_4^{2-}, or H^+ data, and have had to rely on PM_{10} data. Summaries of such PM_{10} epidemiology are shown in Figure 4. There is coherence in the data, as defined by Bates,[2] in terms of the relative risk ratings, with mortality risks increasing from total to cardiovascular to respiratory, and with cough and respiratory conditions being more frequent than mortality.

The results of the carefully controlled prospective cohort studies[6,7] have been analyzed by the US EPA[11] and by Brunekreef,[12] and their analyses indicated that mean lifespan shortening is more than one year, and that the individuals whose lives are shortened are losing about 14 years. An important conclusion that can be drawn from these analyses is that there is considerably more excess

[10] M. Raizenne, L. M. Ware, and F. E. Speizer, *Environ. Health Perspect.*, 1996, **104**, 506.

[11] US EPA, *The Benefits and Costs of the Clean Air Act 1970 to 1990*, US Environmental Protection Agency, Washington, DC, 1997.

[12] B. Brunekreef, *Occup. Environ. Med.*, 1997, **781**.

Figure 6 Plot appearing in PM Staff Paper[5] (Based on data reported by Raizenne *et al.*[10])

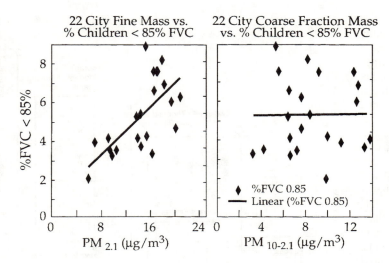

22 City Fine Mass vs. % Children < 85% FVC

22 City Coarse Fraction Mass vs. % Children < 85% FVC

annual mortality associated with chronic fine particle exposure than from the cumulative impacts of daily peaks in exposure on daily mortality.

In the absence of any generally accepted mechanistic basis to account for the epidemiological associations between ambient fine particles on the one hand, and mortality, morbidity, and functional effects on the other, the causal role of PM remains questionable. However, essentially all attempts to discredit the associations on the basis of the effects being due to other environmental variables that may co-vary with PM have been unsuccessful. As shown in Figure 3, the relative risk for daily mortality in relation to PM_{10} is remarkably consistent across communities that vary considerably in their peak concentrations of other criteria air pollutants. The possible confounding influence of adjustments to models to account for weather variables has also been found to be minimal.[13,14]

While mechanistic understanding of processes by which ambient air PM causes human health effects remains quite limited, the credibility of ambient $PM_{2.5}$ as a cause of excess human mortality and morbidity has been enhanced by a series of animal inhalation studies in which rats were exposed to concentrated ambient accumulation mode PM. Godleski *et al.*[15] exposed both healthy and compromised rats to both filtered ambient air and to Boston winter ambient accumulation mode PM that was concentrated $\sim 25 \times$ (*i.e.*, $\sim 250\,\mu g\,m^{-3}$) for 6 h/day for 3 days. For two groups of compromised rats, one with SO_2-induced chronic bronchitis and one with monocrotaline-induced pulmonary hypertension, there was excess mortality during and/or immediately following these exposures. None of the healthy rats died and there was no lung inflammation and only minimal bronchoconstriction following exposure. The hypertensive rats had 19% mortality and evidence of acute inflammation in alveoli and lung interstitium. The bronchitic rats had 37% mortality and responses marked by

[13] J. M. Samet, S. L. Zeger, J. E. Kelsall, J. Xu, and L. S. Kalkstein, *Report on Phase 1.B of the Particle Epidemiology Project*, Health Effects Institute, Cambridge, MA, 1997.

[14] C. A. Pope, III and L. S. Kalkstein, *Environ. Health Perspect.*, 1996, **104**, 414.

[15] J. J. Godleski, C. Sioutas, M. Katler, and P. Koutrakis, *Am. J. Respir. Crit. Care Med.*, 1996, **153**, A15.

airway inflammation, increased mucus, marked bronchoconstriction, interstitial edema, and pulmonary vascular congestion. Compromised rats exposed to filtered ambient air had no comparable responses. Since the animals exposed to filtered air were still exposed to the same pollutant gases (O_3, SO_2, NO_2, CO), the effects are linked to the $PM_{2.5}$. Coarse particles were removed at the inlet to the concentrator, and ultrafine particles ($< 0.15 \, \mu m$) were not concentrated.

In summary, excess daily mortality and morbidity have been related to ambient pollution at current levels in many communities in the US and around the world using available pollutant concentration data. However, it is not yet clear whether any of the pollutant indices used are causally related to the health effects or, if none of them are, which is the best index or surrogate measure of the causal factor(s). This gap can best be addressed by analyses of pollutant associations with mortality and morbidity in locations where a number of different pollutant metrics are available simultaneously, using analytic methods not dependent on arbitrary model assumptions.

PM Exposure Guidelines and Standards

While more research is needed on causal factors for the excess mortality and morbidity associated with PM in ambient air, and on the characterization of susceptibility factors, responsible public health authorities cannot wait for the completion and peer review of this research. It is already clear that the evidence for adverse health effects attributable to PM challenges the conventional paradigm used for setting ambient air standards and guidelines, *i.e.*, that a threshold for adversity can be identified, and a margin of safety can be applied. Excess mortality is clearly an adverse effect, and the epidemiological evidence is consistent with a linear non-threshold response for the population as a whole.

A revision of the Air Quality Guidelines of the World Health Organization-Europe (WHO-EURO) is currently nearing completion. The Working Group of WHO-EURO on PM, at meetings in October 1994 and October 1996 in Bilthoven, The Netherlands, determined that it could not recommend a PM Guideline. Instead, it prepared a tabular presentation of the estimated changes in daily average PM concentrations needed to produce specific percentage changes in: (1) daily mortality; (2) hospital admissions for respiratory conditions; (3) bronchodilator use among asthmatics; (4) symptom exacerbation among asthmatics; and (5) peak expiratory flow. The concentrations needed to produce these changes were expressed in PM_{10} for all five response categories. For mortality and hospital admissions, they were also expressed in terms of $PM_{2.5}$ and SO_4^{2-}. Using this guidance, each national or local authority setting air quality standards can decide how much adversity is acceptable for its population. Making such a choice is indeed a challenge.

In the US, the EPA Administrator promulgated the revised PM NAAQS shown in Table 1 in July 1997 (*Fed. Regis.*, 1997, **62**, 38762–38896) in recognition of the inadequate public health protection provided by enforcement of the 1987 NAAQS for PM_{10}. For PM_{10}, the 50 $\mu g \, m^{-3}$ annual average was retained without change, and the 24-h PM_{10} of 150 $\mu g \, m^{-3}$ was *relaxed* by applying it only to the 99th% value (averaged over 3 years) rather than to the 4th highest over 3 years. These PM_{10} standards were supplemented by the creation of new $PM_{2.5}$ standards. The annual average $PM_{2.5}$ is $15 \mu g\,m^{-3}$, and the 24 hour $PM_{2.5}$ of $65 \mu g\,m^{-3}$ applies to the 98th% value. Implementation of the new $PM_{2.5}$ NAAQS will advance the degree of public health protection for ambient air PM, especially in the eastern US and in some large cities in the west where fine particles are major percentages of PM_{10}.

In the author's view, the new PM NAAQS are not too strict. In terms of its introduction of a more relevant index of exposure and a modest degree of greater public health protection, it represents a prudent judgment call by the Administrator. These NAAQS may not be strict enough to fully protect public health, but there remain significant knowledge gaps on both exposures and the nature and extent of the effects that made the need for more restrictive NAAQS difficult to justify. It is essential that adequate research resources be applied to filling these gaps before the next round of NAAQS revisions during the next first decade of the next century.

3 Ozone

Ozone (O_3) is the indicator for photochemical pollutants. The NAAQS is intended to prevent the health and welfare effects associated with short-term peaks in exposure and to provide protection against more cumulative damage that is suspected, but not clearly established.

Ozone is almost entirely a secondary air pollutant, formed in the atmosphere through a complex photochemical reaction sequence requiring reactive hydrocarbons, nitrogen dioxide (NO_2), and sunlight. It can only be controlled by reducing ambient air concentrations of hydrocarbons, NO_2, or both. Both NO and NO_2 are primary pollutants, known collectively as NO_x. In the atmosphere, NO is gradually converted to NO_2. Motor vehicles, one of the major categories of sources of hydrocarbons and NO_x, have been the target of control efforts, and major reductions ($> 90\%$) have been achieved in the US in hydrocarbon emissions per vehicle. However, there have been major increases in vehicle miles driven. Reductions in NO_x from motor vehicles and stationary-source combustion have been much smaller. The net reduction in exposure has been modest at best, with some reductions in areas with more stringent controls, such as California, and some increases in exposure in other parts of the US. In 1988, there were record high levels of ambient O_3 with exceedance of the former 1-hour maximum 120 ppb limit in 96 communities containing over 150 million people. Since then, there has been a gradual decrease of ambient O_3 in most of the US.

We know a great deal about O_3 chemistry and have developed highly sophisticated O_3 air quality models.[16] Unfortunately, the models, and their applications in control strategies, have clearly been inadequate in terms of community compliance with the NAAQS. We also know a great deal about some of the health effects of O_3. However, much of what we know relates to transient, apparently reversible, effects that follow acute exposures lasting from 5 minutes to 6.6 hours. These effects include changes in lung capacity, flow resistance, epithelial permeability, and reactivity to bronchoactive challenges; such effects

---

[16] J. H. Seinfeld, *J. Air Pollut. Control Assoc.*, 1988, **38**, 616.

can be observed within the first few hours after the start of the exposure and may persist for many hours or days after the exposure ceases. Repetitive daily exposures over several days or weeks can exacerbate and prolong these transient effects. There has been a great deal of controversy about the health significance of such effects and whether such effects are sufficiently adverse to serve as a basis for the O_3 NAAQS.[17–19]

Decrements in respiratory function such as forced vital capacity (FVC) and forced expiratory volume in the first second of a vital capacity maneuver (FEV_1) fall into the category where adversity begins at some specific level of pollutant-associated change. However, there are clear differences of opinion on what the threshold of adversity ought to be. The 1989 Staff Paper[20] included a table in which the responses were categorized as mild, moderate, severe, and incapacitating. The judgement was that mild responses are not adverse, but the other categories were. The 1996 O_3 Staff Paper[21] made numerous elaborations on these gradations, and focused them on persons with impaired respiratory symptoms, as well as on healthy people, because NAAQS are generally set to protect sensitive subgroups of the population. These more elaborate gradations were presented for both healthy persons and persons with impaired respiratory symptoms.

With respect to adversity, the 1996 Staff Paper concluded that responses listed as large or severe were clearly adverse. For responses listed as moderate, it was concluded that they could be considered adverse if there were repetitive exposures.

Review of the Health Effects Literature that Influenced the New O_3 NAAQS

Although we know a great deal about the transient effects on respiratory mechanical function following single exposures to O_3, as shown in Table 4, our current knowledge about the chronic health effects of O_3 is much less complete. The chronic effects include alterations in baseline lung function and structure. Such effects may result from cumulative damage and/or from the side effects of adaptive responses to repetitive daily or intermittent exposures.

In terms of functional effects, we know that single O_3 exposures to healthy nonsmoking young adults at concentrations in the range of 80–200 ppb produce a complex array of pulmonary responses including decreases in respiratory function and athletic performance, and increases in symptoms, airway reactivity, neutrophil content in lung lavage, and rate of mucociliary particle clearance. The respiratory function responses to O_3 in purified air in chambers that occur at concentrations of 80 or 100 ppb when the exposures involve moderate exercise over 6 h or more are illustrated in Figure 7. Comparable responses require

[17] M. Lippmann, *J. Air Pollut. Control Assoc.*, 1988, **38**, 881.

[18] M. Lippmann, in *The Handbook of Environmental Chemistry, Volume 4: Part C: Air Pollution*, ed. O. Hutzinger, Springer, Heidelberg, 1991, p. 31.

[19] M. Lippmann, *J. Exp. Anal. Environ. Epidemiol.*, 1993, **3**, 103.

[20] US EPA, *Review of the National Ambient Air Quality Standards for Ozone—Assessment of Scientific and Technical Information—OAQPS Staff Paper*, EPA-450/2-92/001, NTIS, Springfield, VA, 1989.

[21] US EPA, *Review of National Ambient Air Quality Standards for Ozone—Assessment of Scientific and Technical Information—OAQPS Staff Paper*, EPA-452/A-96-007, US EPA-OAQPS, Research Triangle Park, NC, 1996.

Table 4 Mean functional changes per part per billion ozone after moderate or heavy exercise: comparison of results from field and chamber exposure studies

Investigator Subjects, age (yr)	Minute ventilation (L)	Exposure (exercise) time (min)	Ozone concentration (ppb)	Mean rate of functional change FVC [mL (ppb)$^{-1}$]	FEV_1 [mL (ppb)$^{-1}$]	$FEF_{25-75\%}$ [mL s^{-1} (ppb)$^{-1}$]
Folinsbee (1988) 10 M, 18–33	40	395 (300)	120[c]	−3.8	−4.5	−5.0
McDonnell (1983) 22 M, 22.3 H 3.1[a]	65	120 (60)	120[c]	−1.4	−1.3	−2.9
20 M, 23.3 H 2[a]	65	120 (60)	180[c]	−1.8	−1.6	−3.0
Kulle (1985) 20M, 25.3 H 4.1[a]	68	120 (60)	150[c]	−0.5	−0.2	−2.1
Linn (1980) 24 M, 18–33	68	120 (60)	160[c]	−0.7	−0.6	−1.1
Spektor (1988b) 1 M, 9 F, 28–44	38.4 ± 12.3[a]	34.4 ± 9.9[a]	21–124[b]	−1.9	−1.8	−6.7
7 M, 3 F, 22–40	64.6 ± 10.0[a]	26.7 ± 8.7[a]	21–124[b]	−2.9	−3.0	−9.7
Spektor (1988a) 53 M, 38 F, 7–13	—	150–550	19–113[b]	−1.0	−1.4	−2.5
Avol (1987) 33 M, 33 F, 8–11	22	60 (60)	113[b]	−0.3	−0.3	—
Avol (1985) 46 M, 13 F, 12–15	32	60 (60)	150[c]	−0.7	−0.8	−0.7
McDonnell (1985b) 23 M, 8–11	39	150 (60)	120[c]	−0.3	−0.5	−0.6

[a] Mean ± SD. [b] Ozone concentration within ambient mixture. [c] Ozone concentration within purified air.
From M. Lippmann, in *Environmental Toxicants*, ed. M. Lippmann, Van Nostrand Reinhold, 1992, pp. 465–519.

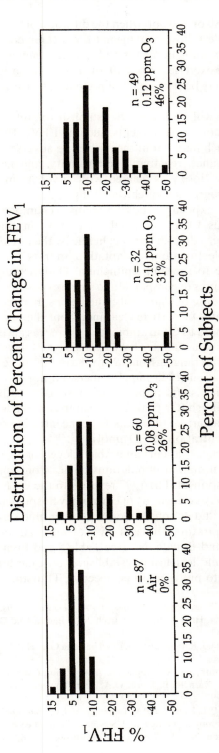

Figure 7 Percentage change in FEV_1 in healthy nonsmokers following 6.6 h exposures to clean air and O_3 at 80, 100, and 120 ppb during exercise lasting 50 minutes of each hour for studies performed at EPA's Clinical Research Laboratory at Chapel Hill, NC. Each box shows the number of subjects studied and the percentage of subjects with reductions in FEV_1 that were greater than 10% (Adapted from ref. 41)

concentrations of 180 or 200 ppb when the duration of exposure is 2 h or less. On the other hand, Table 4 shows that mean FEV_1 decrements $> 5\%$ have been seen at 100 ppb of O_3 in ambient air for children at summer camps and for adults engaged in outdoor exercise for only 0.5 h. The apparently greater responses to O_3 in ambient air may be related to the presence of, or prior exposures to, acidic aerosol.

Further research will be needed to establish the interrelationships between small transient functional decrements, such as FEV_1, PEFR, and mucociliary clearance rates, which may not in themselves be adverse effects, and changes in symptoms, performance, reactivity, permeability, and neutrophil counts. The latter may be more closely associated with adversity in themselves or in the accumulation or progression of chronic lung damage.

Successive days of exposure of adult humans in chambers to O_3 at current high ambient levels leads to a functional adaptation in that the responses are attenuated by the third day, and are negligible by the fifth day.[22,23] On the other hand, a comparable functional adaptation in rats[24] does not prevent the progressive damage to the lung epithelium. Daily exposures of animals also increase other responses in comparison to single exposures, such as a loss of cilia, a hypertrophic response of Clara cells, alterations in macrophage function, and alterations in the rates of particle clearance from the lungs.

For children exposed to O_3 in ambient air there was a week-long baseline shift in peak flow following a summer haze exposure of 4 days' duration, with daily peak O_3 concentrations ranging from 125 to 185 ppb.[25] Since higher concentrations used in adult adaptation studies in chambers did not produce such effects, it is possible that baseline shifts require the presence of other pollutants in the ambient air. A baseline shift in peak flow in camp children was also seen following a brief episode characterized by a peak O_3 concentration of 143 ppb and a peak acidic aerosol concentration of 559 nmol m^{-3}.[26]

The clearest evidence that current ambient levels of O_3 are closely associated with health effects in human populations comes from epidemiological studies focused on acute responses. The 1997 revision to the O_3 NAAQS relied heavily for its quantitative basis on a study of emergency hospital admissions for asthma in New York City[27] and its consistency with other time-series studies of hospital admissions for respiratory diseases in Toronto, all of Southern Ontario, Montreal, Detroit, and Buffalo, NY (see Table 5 and Figure 8). However, other acute responses, while less firmly established on quantitative bases, are also occurring. In order to put them in perspective, Thurston[28] prepared a graphic

[22] S. M. Horvath, J. A. Gliner, and L. J. Folinsbee, *Am. Rev. Respir. Dis.*, 1981, **123**, 496.

[23] T. J. Kulle, L. R. Sauder, H. D. Kerr, B. P. Farrell, M. S. Bermel, and D. M. Smith, *Am. Ind. Hyg. Assoc. J.*, 1982, **43**, 832.

[24] J. S. Tepper, D. L. Costa, J. R. Lehmann, M. F. Weber, and G. E. Hatch, *Am. Rev. Respir. Dis.*, 1989, **140**, 493.

[25] P. J. Lioy, T. A. Vollmuth, and M. Lippmann, *J. Air Pollut. Control Assoc.*, 1985, **35**, 1068.

[26] M. E. Raizenne, R. T. Burnett, B. Stern, C. A. Franklin, and J. D. Spenger, *Environ. Health Perspect.*, 1989, **79**, 179.

[27] G. D. Thurston, K. Ito, P. L. Kinney, and M. Lippmann, *J. Exp. Anal. Environ. Epidemiol.*, 1992, **2**, 429.

[28] G. D. Thurston, testimony submitted to US Senate Committee on Environment and Public Works, Subcommittee on Clean Air, Wetlands, Private Property, and Nuclear Safety, February 1977.

Table 5 Summary of effect estimates for ozone in recent studies of respiratory hospital admissions

Location	Reference	Respiratory admission category	Effect size ($\pm$ SE) (admissions/100 ppb O_3/ day/10^6 persons)	Relative risk (95% CI)[a] (RR of 100 ppb O_3, 1-h max)
New York City, NY[b]	Thurston et al., (1992)	All	1.4 ($\pm$ 0.5)	1.14 (1.06–1.22)
Buffalo, NY[b]	Thurston et al., (1992)	All	3.1 ($\pm$ 1.6)	1.25 (1.04–1.46)
Ontario, Canada[b]	Burnett et al., (1994)	All	1.4 ($\pm$ 0.3)	1.10 (1.06–1.14)
Toronto, Canada[b]	Thurston et al., (1994)	All	2.1 ($\pm$ 0.8)	1.36 (1.13–1.59)
Montreal, Canada[c]	Delfino et al., (1994a)	All	1.4 ($\pm$ 0.5)	1.22 (1.09–1.35)
Birmingham, AL[d]	Schwartz (1994a)	Pneumonia in elderly	0.73 ($\pm$ 0.54)	1.11 (0.97–1.26)
Birmingham, AL[d]	Schwartz (1994a)	COPD in elderly	0.83 ($\pm$ 0.33)	1.13 (0.92–1.39)
Detroit, MI[d]	Schwartz (1994b)	Pneumonia in elderly	0.82 ($\pm$ 0.26)	1.22 (1.12–1.35)
Detroit, MI[d]	Schwartz (1994b)	COPD in elderly	0.90 ($\pm$ 0.41)	1.25 (1.07–1.45)
Minneapolis, MD[d]	Schwartz (1994c)	Pneumonia in elderly	0.41 ($\pm$ 0.19)	1.117 (1.03–1.39)
Minneapolis, MN[d]	Schwartz (1994c)	COPD in elderly	—[e]	—[e]

[a]One-way ($\beta \pm 1.65$ SE). [b]1-h daily maximum ozone data employed in analysis. [c]8-h daily maximum ozone data employed in analysis.
[d]24-h daily average ozone data employed in analysis (1 h/24 h av. ratio = 2.5 assumed to compute effects and RR estimates).
[e]Not reported (nonsignificant).
From *Review of National Ambient Air Quality Standards for Ozone*, OAQPS Staff Paper, EPA-452/R-96-007, US EPA, Research Triangle Park, NC, 1996.

Figure 8 Average number of adjusted respiratory admissions among all 168 hospitals in Lower Ontario by decile of the daily 1-h maximum O_3 level (ppm), lagged 1 day (Adapted from ref. 41)

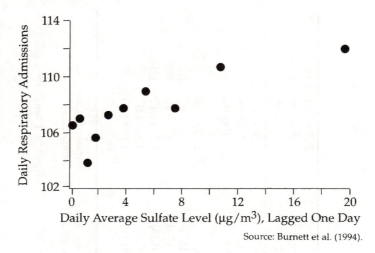

Source: Burnett et al. (1994).

Figure 9 Pyramid summarizing the adverse effects of ambient O_3 in New York City that can be averted by reduction of mid-1990s levels to those meeting the 1997 NAAQS revision (Data assembled by Dr. G. D. Thurston for testimony to US Senate Committee on Public Works)

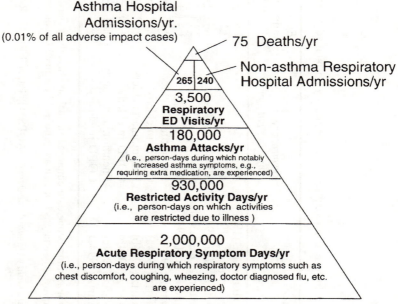

presentation showing the extent of related human responses based on the exposure–response relationships established in a variety of studies reviewed earlier in this chapter. This is shown, for New York City, in Figure 9. It estimates the extent of the human health responses to ambient ozone exposures in New York City that could be avoided by full implementation of the new O_3 NAAQS of 80 ppb averaged over 8 h. The extent of effects avoided on a national scale would be much larger. However, they have not been estimated, and would require knowledge of current (1997) O_3 levels and populations at risk in other parts of the country.

Chronic human exposures to ambient air appear to produce a functional adaptation that persists for at least a few months after the end of the O_3 season but dissipates by the spring.[29] Several population-based studies of lung function[30,31] indicate that there may be an accelerated aging of the lung associated with living in communities with persistently elevated ambient O_3, but the limited ability to accurately assign exposure classifications of the various populations in these studies makes a cautious assessment of these provocative data prudent.

Some evidence for chronic effects of O_3 were reported from an analysis of pulmonary function data in a national population study in 1976–80, *i.e.*, the second National Health and Nutrition Examination Survey (NHANES II).[32] Using ambient O_3 data from nearby monitoring sites, Schwartz reported a highly significant O_3-associated reduction in lung function for people living in areas where the annual average O_3 concentrations exceeded 40 ppb.

An autopsy study of 107 lungs from 14–25 year old accident victims in Los Angeles County by Sherwin and Richters[33] reported that 27% had what were judged to be severe degrees of structural abnormalities and bronchiolitis not expected for such young subjects, while another 48% of them had similar, but less severe, abnormalities. In the absence of corresponding analyses of lungs of comparable subjects from communities having much lower levels of air pollution, the possible association of the observed abnormalities with chronic O_3 exposure remains speculative. Some of the abnormalities observed could have been due to smoking and/or drug abuse, although the authors noted that published work on the association between smoking and small airway effects showed lesser degrees of abnormality.[34]

Although the results of these epidemiological and autopsy studies are strongly suggestive of serious health effects, they have been found wanting as a basis for standards setting by EPA staff. The basis for skepticism lies in the uncertainty about the exposure characterization of the populations and the lack of control of possibly important confounding factors. Some of these limitations are inherent in large-scale epidemiological studies. Others can be addressed in more carefully focused study protocols.

The plausibility of accelerated aging of the human lung from chronic O_3 exposure is greatly enhanced by the results of chronic animal exposure studies at near ambient O_3 concentrations in rats and monkeys.[35–39] There is little reason

[29] W. S. Linn, E. L. Avol, D. A. Shamoo, R. C. Peng, L. M. Valencia, D. E. Little, and J. D. Hackney, *Toxicol. Ind. Health*, 1988, **4**, 505.

[30] R. Detels, D. P. Tashkin, J. W. Sayre, S. N. Rokaw, A. H. Coulson, F. J. Massey, and D. H. Wegman, *Chest*, 1987, **92**, 594.

[31] K. H. Kilburn, R. Warshaw, and J. C. Thornton, *Am. J. Med.*, 1985, **79**, 23.

[32] J. Schwartz, *Environ. Res.*, 1989, **50**, 309.

[33] R. P. Sherwin and V. Richters, in *Tropospheric Ozone and the Environment (TR-19)*, ed. R. L. Berglund, D. R. Lawson, and D. J. McKee, Air & Waste Management Assoc., Pittsburgh, 1991, p. 178.

[34] M. G. Cosio, K. A. Hole, and D. E. Niewohner, *Am. Rev. Respir. Dis.*, 1980, **122**, 265.

[35] W. S. Tyler, N. K. Tyler, J. A. Last, M. J. Gillespie, and T. J. Barstow, *Toxicology*, 1988, **50**, 131.

[36] D. M. Hyde, C. G. Plopper, J. R. Harkema, J. A. St. George, W. S. Tyler, and D. L. Dungworth, in *Atmospheric Ozone Research and Its Policy Implications*, ed. T. Schneider, S. D. Lee, G. J. R. Wolters, and L. D. Grant, Elsevier, Nijmegen, The Netherlands, 1989.

[37] Y. Huang, L. Y. Chang, F. J. Miller, J. A. Graham, J. J. Ospital, and J. D. Crapo, *Am. J. Aerosol Med.*, 1988, **1**, 180.

to expect humans to be less sensitive than rats or monkeys. On the contrary, humans have a greater dosage delivered to the respiratory acinus than do rats for the same exposures. Another factor is that the rat and monkey exposures were to confined animals with little opportunity for heavy exercise. Thus humans who are active outdoors during the warmer months may have greater effective O_3 exposures than the test animals. Finally, humans are exposed to O_3 in ambient mixtures. The potentiation of the characteristic O_3 responses by other ambient air constituents seen in the short-term exposure studies in humans and animals may also contribute toward the accumulation of chronic lung damage from long-term exposures to ambient air containing O_3.

The lack of a more definitive database on the chronic effects of ambient O_3 exposures on humans is a serious failing that must be addressed with a long-term research program. The potential impacts of such exposures on public health deserve serious scrutiny and, if they turn out to be substantial, strong corrective action. Further controls on ambient O_3 exposure will be extraordinarily expensive and will need to be very well justified.

Ozone Exposure Standards

The US Occupational Safety and Health Administration's (OSHA) permissible exposure limit (PEL) for O_3 is 100 parts per billion (ppb), equivalent to $235 \mu g \, m^{-3}$, as a time-weighted average for 8 h/day, along with a short-term exposure limit of 300 ppb for 15 min.[40] The American Conference of Governmental Industrial Hygienists[41] threshold limit value (TLV) for occupational O_3 exposure is 100 ppb as an 8-hour time-weighted average for light work, 80 ppb for moderate work, and 50 ppb for heavy work.

The initial primary (health-based) NAAQS established by the EPA in 1971 was 0.80 ppb of total oxidant as a 1-hour maximum not to be exceeded more than once per year. The NAAQS was revised in 1979 to 120 ppb of O_3 as a 1-hour maximum not to be exceeded more than four times in three years.

EPA initiated a review of the 1979 NAAQS in 1983, and completed a Criteria Document for Ozone in 1986 and updated it in 1992.[42] However, the Agency did not decide either to retain the 1979 standard or to promulgate a new one until it was compelled to do so by a August 1992 court order. In response, the EPA decided, in March 1993, to maintain the existing standard and to proceed as rapidly as possible with the next round of review. This expedited review was

[38] J. S. Tepper, M. J. Wiester, M. F. Weber, S. Fitzgerald, and D. L. Costa, *Fundam. Appl. Toxicol.*, 1991, **17**, 52.

[39] L.-Y. Chang, Y. Huang, B. L. Stockstill, J. A. Graham, E. C. Grose, M. G. Menache, F. J. Miller, D. L. Costa, and J. D. Crapo, *Toxicol. Appl. Pharmacol.*, 1992, **115**, 241.

[40] US DOL, *Fed. Regist.*, 1989, **54**, 2332.

[41] ACGIH, *Threshold Limit Values and Biological Exposure Indices for 1997*, American Conference of Governmental Industrial Hygienists, Cincinnati, 1997.

[42] US EPA, *Summary of Selected New Information on Effects of Ozone on Health and Vegetation: Supplement to 1986 Air Quality Criteria for Ozone and Other Photochemical Oxidants*, EPA/600/8-88/105F, ECAO, NTIS, Springfield, VA, 1992.

completed with the publication of both a new criteria document[43] and staff paper[21] in 1996. In July 1997 the EPA Administrator promulgated a revised primary O_3 NAAQS of 80 ppb as an 8-hour time-weighted average daily maximum, with no more than four annual exceedances, and averaged over three years (*Fed. Regist.*, 1997, **62**, 38762–38896) (see Table 1). The reason for the switch from one allowable annual exceedance to four was to minimize the designation of NAAQS non-attainment in a community that was triggered by rare meteorological conditions especially conducive to O_3 formation. The goal was to have a more stable NAAQS that allowed for extremes of annual variations of weather. The switch to an 8-hour averaging time was in recognition that ambient O_3 in much of the US has broad daily peaks, and that human responses are more closely related to total daily exposure than to brief peaks of O_3 exposure. Since the 120 ppb, 1-h average, one exceedance NAAQS was approximately equivalent to an 8-h average, four exceedance NAAQS at a concentration a little below 90 ppb in average stringency in the US as a whole, the 1997 NAAQS represents about a 10% reduction in permissible O_3 exposure. While this new NAAQS will be difficult to achieve in much of the southern and eastern US, it represents a prudent public health choice considering the extent and severity of the effects that O_3 produces in sensitive segments of the population.

The effects of concern with respect to acute response in the population at large are reductions in lung function and increases in respiratory symptoms, airway reactivity, airway permeability, and airway inflammation. For persons with asthma, there are also increased rates of medication usage and restricted activities. Margin-of-safety considerations included: (1) the influence of repetitive elicitation of such responses in the progression of chronic damage to the lung of the kinds seen in chronic exposure studies in rats and monkeys; and (2) evidence from laboratory and field studies that ambient air co-pollutants potentiate the responses to O_3.

4 Acknowledgements

This research is part of a Center program supported by Grant ES 00260 from the National Institute of Environmental Health Sciences. It has been based, in part, on material contained in the chapters on ambient particulate matter and ozone written by the author for the second edition of *Environmental Toxicants—Human Exposures and Their Health Effects* being published in 1998.

[43] US EPA, *Air Quality Criteria for Ozone and Related Photochemical Oxidants*, EPA/600/P-93/004F, US EPA, National Center for Environmental Assessment, Research Triangle Park, NC, 1996.

Health Effects of Indoor Air Pollutants

PAUL T. C. HARRISON

1 Introduction

Much attention is paid to outdoor air quality and its impact on health, but individuals may spend 90% or more of their time indoors. For many people, especially potentially vulnerable groups such as the very young, the elderly and the sick, this means at home. Hence the quality of air inside the home environment is extremely important, and although a good deal of public interest and concern continues to be directed at the effects of outdoor air pollution, there is a growing tide of scientific opinion that the quality of air in the home environment is of equal or greater significance to human health and well-being. This growing interest has resulted in increasing research activities and, importantly, heightened awareness among regulators and policy makers.

This article reviews current information on levels and the risks to health and well-being of some major indoor air pollutants in dwellings. Because indoor air quality is very much dependent on prevailing climate, day-length, building construction, use of different fuels and heating/cooking methods, *etc.*, assessments of this kind tend, by necessity, to have a national or regional focus and to make most use of 'local' exposure measurements. Therefore this paper concentrates particularly on the UK situation, whilst acknowledging and drawing information from important studies and developments elsewhere.[1] Many of the considerations and conclusions will be relevant to other countries, especially those with similar climate and building stock.

For a number of pollutants found indoors, the main sources are outside. Where significant indoor sources exist, these will tend to dominate personal exposure. Certainly it is known that the behaviour of individuals and their activity patterns (reflecting the time spent in various different 'micro-environments') can markedly affect their exposure to a range of air pollutants. Increasingly, therefore, the need is being recognized to take much better account of indoor exposures and to understand the importance of personal behaviour patterns. Only in this way can adequate assessments be made of the true impact of air pollution on health.

There are other important factors which affect how indoor air pollution is

---

[1] IEH, *Assessment on Indoor Air Quality in the Home*, Assessment A2, Institute for Environment and Health, Leicester, 1996.

considered in relation to outdoor air pollution. One of these is risk perception. For many individuals, for whatever reason, the perception of risk from outdoor air is substantially higher than for indoor air; indeed, it is unusual for the home environment to be considered 'hazardous' in this sense. Another important factor is 'controllability'. The principal outdoor sources of pollution (vehicles, factories, *etc.*) lend themselves to formal legislative control procedures, whereas an individual's exposure in the home is very much dominated by personal choice and behaviour with respect to ventilation, use of personal and consumer products, *etc.* Nevertheless, there is scope for control of emissions, especially from appliances and building products, and this is one area where efforts are currently under way (see below).

2   The Pollutants

A large number of natural and man-made substances can be identified in the air inside a typical home, many of which arise from sources within the home. The impetus over recent years to conserve energy has resulted in warmer, 'tighter' buildings with much reduced air exchange and therefore a greater propensity for indoor pollutants to build up. The combination of reduced ventilation rates (especially in winter), warmer and more humid conditions indoors, together with the greater use and diversity of materials, furnishings and consumer products, has resulted in concentrations of a wide range of pollutants occurring indoors at levels exceeding those outdoors. The following sections review exposure and health data for some of the most important indoor pollutants.

Nitrogen Dioxide (NO₂)

Nitrogen dioxide, a product of fossil fuel combustion, is one of the most ubiquitous indoor pollutants, especially in homes with gas cookers and other unflued combustion appliances. Because of this widespread exposure it has been the focus of much attention with respect to possible health effects. It is of particular importance in the UK, where almost 50% of homes are equipped with gas cookers and thus approximately 30 million people are potentially exposed indoors to NO_2 and related products of combustion.

Long-term average outdoor NO_2 levels in the UK fluctuate with season and degree of urbanization. They are typically well below WHO guideline values, although short-term peaks (*e.g.* one hour) can reach high levels, particularly in areas with heavy road traffic.[2] Outdoor levels are important determinants of indoor levels, but the latter are normally lower unless there is an indoor source.[3]

[2] MAAPE, *Oxides of Nitrogen*, Advisory Group on the Medical Aspects of Air Pollution Episodes, Third Report, HMSO, London, 1993.
[3] S. K. D. Coward and G. J. Raw, in *Indoor Air Quality in Homes: Part 1, The Building Research Establishment Indoor Environment Study*, ed. R. W. Berry, S. K. D. Coward, D. R. Crump, M. Gavin, C. P. Grimes, D. F. Higham, A. V. Hull, C. A. Hunter, I. G. Jeffrey, R. G. Lea, J. W. Llewellyn and G. J. Raw, Construction Research Communications, London, 1996, p. 67.

Recent studies in UK homes with and without gas cookers[3-6] have shown one- or two-week averages ranging from 25 to 70 μg m^{-3} and from 13 to 40 μg m^{-3}, respectively. Continuous monitoring in kitchens with gas cookers[6] has shown one-hour average levels of up to 1115 μg m^{-3}; this compares with the 1987 WHO one-hour guideline value of 400 μg m^{-3} (soon to be reduced to 200 μg m^{-3}). These limited data suggest that in many homes using gas for cooking, levels in the kitchen (and possibly in other rooms) approach or exceed this guideline value.

The most frequent end-points in studies looking at health effects from indoor NO$_2$ have been respiratory illness and/or symptoms in children. These outcomes have been defined differently in different investigations, and may not all represent the same disease process. For example, symptoms such as wheeze and cough may indicate a chronic disorder such as asthma or an acute infection in an otherwise normal person. However, many of the published reports do not discriminate clearly between infections and other types of respiratory disease, which must therefore be considered together. Eleven of the epidemiological studies looking at respiratory illness in children were included in a meta-analysis carried out by Hasselblad *et al.*[7] Four different statistical methods were used to combine the results of the studies and calculate summary odds ratios. All four methods produced the same estimate for the effects of NO$_2$ exposure on respiratory illness, and there was little or no change in the odds ratio when the analysis was limited to studies of children aged 5–12 years, with studies analysed separately according to whether NO$_2$ was measured directly or inferred from the presence of a gas cooker. The authors concluded that children exposed to a long-term increase of 30 μg m^{-3} NO$_2$ (equivalent to having a gas cooker) suffer a 20% increase in their risk of respiratory illness. Other studies, not included in Hasselblad's analysis, have produced inconsistent results. Two of these were part of the US Six Cities studies,[8,9] and only showed associations of gas cooking with previous respiratory disease (before age two). Also in the US, an early study showed an association of gas cookers with cough, but associations with other symptoms, although positive, were not statistically significant.[10] In the Netherlands, one study found that use of unvented water heaters and increased personal exposure to NO$_2$ were associated with a higher prevalence of respiratory symptoms,[11] while in another investigation, indoor levels of NO$_2$ did not differ between children with respiratory problems and asymptomatic controls.[12] A Canadian case-control study found higher levels of NO$_2$ personal exposure among asthmatic children

[4] R. J. W. Melia, S. Chinn and R. J. Rona, *Atmos. Environ.*, 1990, **24B**, 177.

[5] G. J. Raw and S. K. D. Coward, in *Proceedings of Unhealthy Housing: The Public Health Response*, University of Warwick, Coventry, 1992.

[6] D. Ross, *Continuous and Passive Monitoring of Nitrogen Dioxide in UK Homes*, BRE Note N109/94, Buildings Research Establishment, Watford, 1994.

[7] V. Hasselblad, D. M. Eddy and D. J. Kotchman, *J. Air Waste Manage. Assoc.*, 1992, **42**, 662.

[8] F. E. Speizer, B. Ferris, Y. M. M. Bishop and J. Spengler, *Am. Rev. Respir. Dis.*, 1980, **121**, 3.

[9] D. W. Dockery, J. D. Spengler, L. M. Neas, F. E. Speizer, B. G. Ferris, J. H. Ware and B. Brunekreef, in *Air Waste Management Transaction Series: TR-15,* ed. J. Harper, Air & Waste Management Assoc., Pittsburgh, 1989, p. 262.

[10] R. Dodge, *Arch. Environ. Health*, 1982, **37**, 151.

[11] D. Houthuijs, B. Remijn, B. Brunekreef and R. de Konig, in *Proceedings of Indoor Air '87*, Institute for Water, Soil and Air Hygiene, Berlin, 1987, p. 463.

[12] G. Hoek, B. Brunekreef, R. Meijer, A. Scholten and J. Boleij, *Int. Arch. Occup. Health*, 1984, **55**, 79.

than among non-asthmatics.[13] Of the studies that have looked specifically at health effects in infants or neonates, no relationship has been shown between respiratory illness and the presence of a gas cooker[14] or NO_2 levels in the bedroom or kitchen.[15] In adults, studies of the relation between the use of gas cookers and the occurrence of respiratory illness have provided conflicting results. Two studies, for example, showed no significant association; indeed, the subjects using gas had slightly less respiratory illness than those who used electricity for cooking.[16,17] In a third study there was some association between gas cookers and breathlessness in non-smoking men but not in women, whose exposure to NO_2 might be expected to be higher.[18] In another study, prevalence of respiratory symptoms in women was associated with the frequency with which the kitchen was filled with heavy cooking fumes but not with gas cooking *per se*.[19] However, a more recent paper has shown an apparent connection between gas cooking and ill-health in women.[20] Thus there is little consistent evidence at present to suggest that the use of gas cookers has any important effect on the incidence of respiratory illness in adults, although further work is needed.

A number of studies have looked at the potential influence of gas cooking and/or indoor NO_2 levels on measures of pulmonary function rather than clinical illness. In children, some studies have reported small negative effects of gas over electric cooking on spirometric indices, but between the different indices the effects are generally inconsistent. In adults, small detrimental effects of gas cooking have been reported on one or more measures of pulmonary function,[20] but again there are inconsistencies between indices. While some of these studies included NO_2 measurements, because of possible confounding it is not possible to attribute any differences in lung function to indoor NO_2 alone, even where associations were found. In support of this, single and repeated chamber studies of various durations have failed to show any effect of NO_2 up to $1880 \, \mu g \, m^{-3}$ on indices of lung function.[2] A few studies have examined the effect of cooking or NO_2 levels on pulmonary function in patients with asthma, but the numbers are generally too small to permit firm conclusions to be drawn.

Conclusions. Overall, the published evidence on health effects of NO_2 points most to a hazard of respiratory illness in children, perhaps resulting from increased susceptibility to infection. However, in interpreting this evidence, several sources of uncertainty should be taken into account, including publication bias, reporting bias, multiple testing errors, confounding, pollutant interactions and use of proxy measures of NO_2 exposure. Also there is a dearth of studies on asthmatics, bronchitics and other potentially susceptible groups. A number of

[13] C. Infante-Rivard, *Am. J. Epidemiol.*, 1993, **137**, 834.

[14] S. A. Ogston, C. V. du Florey and C. H. M. Walker, *Br. Med. J.*, 1985, **290**, 957.

[15] J. M. Samet, W. E. Lambert, B. J. Skipper, B. J. Cushing, W. C. Hunt, S. A. Young, L. C. McLaren, M. Schwab and J. D. Spengler, *Nitrogen Dioxide and Respiratory Illness in Children. Part I: Health Outcomes*, Research Report Number 58, Health Effects Institute, Cambridge, 1993.

[16] M. D. Keller, R. R. Lanese, R. I. Mitchell and R. W. Cote, *Environ. Res.*, 1979, **19**, 495.

[17] M. D. Keller, R. R. Lanese, R. I. Mitchell and R. W. Cote, *Environ. Res.*, 1979, **19**, 504.

[18] G. W. Comstock, M. B. Meyer, K. J. Helsing and M. S. Tockman, *Am. Rev. Respir. Dis.*, 1981, **124**, 143.

[19] T. P. Ng, K. P. Hui and W. C. Tan, *J. Epidemiol. Community Health*, 1993, **47**, 454.

[20] D. Jarvis, S. Chinn, C. Luczynska and P. Burney, *Lancet*, 1996, **347**, 426.

research needs are apparent, including: the identification of homes with high indoor levels of NO_2 to establish whether such levels are associated with detectable health effects; further work on health effects of NO_2 and gas cooking in potentially susceptible groups; identification of reasons why some homes have high levels of NO_2; and further information on the pathogenic mechanisms of NO_2 toxicity in the lung. While it does seem clear that any risk of respiratory illness from the levels of NO_2 currently found in most homes is small, it would seem prudent to encourage any measure that will minimize indoor NO_2 levels. This is especially relevant because of the large number of people potentially exposed and because of the uncertainties regarding effects on susceptible groups such as asthmatics and bronchitics and people who spend a particularly large proportion of time indoors such as very young infants and the elderly. Further work is needed on exposure to gas combustion products (the full mixture) and their effects on health, including the postulated potentiation of responses to indoor allergens, such as those from house dust mites, by concomitant exposure to irritant gases such as NO_2.

Formaldehyde and Other Volatile Organic Compounds (VOCs)

VOCs originate from a number of sources within the indoor environment, including building materials, paints, furnishings, furniture, adhesives, cleaning agents, tobacco smoke and the occupants themselves. Numerous VOCs, representing most organic families (typically aliphatic and aromatic hydrocarbons, halogenated compounds and aldehydes) have been measured in indoor air. Formaldehyde occurs ubiquitously in the environment. It is produced naturally and by many industrial processes, and is found in vehicle exhausts and cigarette smoke. It also occurs naturally in fruits and vegetables and other foods. In residential indoor air, the principal source of formaldehyde is off-gassing from urea formaldehyde foam insulation (UFFI) and particle board used in construction; other sources are furniture, furnishing and household cleaning agents. For both formaldehyde and VOCs, airborne concentrations depend on the age of the source material and ventilation, temperature and humidity. Active and passive cigarette smoking also contribute to total exposure. Exposure to these substances is therefore widespread and the potential consequences to health are significant.

Formaldehyde.   Within the UK, the most extensive investigation of formaldehyde exposure in the home has been that conducted by the Building Research Establishment (BRE) in 180 homes mainly within the Avon area.[21] These studies demonstrated somewhat increased formaldehyde levels in newer homes, homes with integral garages, homes with new furnishings and recently decorated homes. Outdoor levels in the BRE study were around one tenth of those found indoors; most other investigations have similarly shown outdoor levels of formaldehyde to be lower than indoor levels. The mean annual indoor formaldehyde levels

---

[21] V. M. Brown, D. R. Crump and M. Gavin, in *Indoor Air Quality in Homes: Part 1, The Building Research Establishment Indoor Environment Study*, ed. R. W. Berry, S. K. D. Coward, D. R. Crump, M. Gavin, C. P. Grimes, D. F. Higham, A. V. Hull, C. A. Hunter, I. G. Jeffrey, R. G. Lea, J. W. Llewellyn and G. J. Raw, Construction Research Communications, London, 1996, p. 18.

found (0.020–0.025 mg m^{-3}, according to room sampled) were at the lower end of the 0.01–0.1 mg m^{-3} range reported in a European survey of formaldehyde concentrations in residential homes and schools[22] (some with UFFI, others without), and were similar to or less than those reported in contemporary US studies. Overall, the homes in the UK BRE survey do not appear to differ greatly, in terms of formaldehyde levels, from residential dwellings elsewhere in the world; certainly there is no evidence that concentrations are greater. Many studies on formaldehyde levels in homes were carried out in the USA in the 1980s, in both conventional and mobile homes. The latter contained many formaldehyde-emitting materials such as UFFI and were considered to pose potential health problems. A review on indoor air pollution by Samet *et al.*[23] indicated that in homes with UFFI, formaldehyde concentrations were about twice those in homes without UFFI (0.02–0.16 and 0.04–0.08 mg m^{-3}, respectively). North American studies have, like the UK BRE investigations, demonstrated higher formaldehyde concentrations in newer compared with older dwellings. In a study of Tennessee homes,[24] for example, those less than five years old had mean formaldehyde levels of around 0.1 mg m^{-3}, whereas in homes between five and fifteen years old the mean was 0.05 mg m^{-3} and in older homes the mean concentration was 0.038 mg m^{-3}. The half-life of formaldehyde for new homes appears to be around four to five years.[25] Energy conservation measures in dwellings have been shown to cause an increase in exposure to formaldehyde. In one US study,[26] conventional houses had a mean formaldehyde level of 0.05 mg m^{-3} compared with 0.08 mg m^{-3} in energy-efficient houses; for normal *versus* energy-efficient condominiums, the level was 0.11 mg m^{-3} compared with 0.22 mg m^{-3}.

Numerous studies, including those in occupational settings, have investigated and reported health effects related to exposure to formaldehyde by inhalation. Effects range from subtle neuropsychological changes, mucous membrane irritation of the eyes, nose and throat, and airway irritation, to asthma and cancer. The odour threshold for formaldehyde is in the range 0.06–1.2 mg m^{-3} and for throat irritation is in the range 0.12–3.0 mg m^{-3} for most individuals. Eye irritation has been reported at levels as low as 0.01 mg m^{-3}. In a review of a number of US studies in which symptoms among residents of mobile homes or homes with UFFI had been investigated,[27] exposures ranged from 0 to 9.6 mg m^{-3}. Although the studies are suggestive of some irritant effects, the limited exposure reporting and inconsistent symptoms reporting in these studies do not allow conclusions regarding specific effects at particular levels of exposure. It is, however, clear that mucous membrane irritation to the eyes and throat can

[22] ECA-IAQ, *Indoor Air Pollution by Formaldehyde in European Countries*, European Collaborative Action 'Indoor Air Quality and its Impact on Man', Report No 7. EUR 13216 EN, Office of Publications for the European Communities, Luxembourg, 1990.

[23] J.M. Samet, M.C. Marbury and J.D. Spengler, *Am. Rev. Respir. Dis.*, 1987, **136**, 1486.

[24] A.R. Hawthorne, R.B. Gammage, C.S. Dudney, D.R. Womack, S.A. Morris, R.R. Westley and K.C. Gupta, in *Specialty Conference on Measurement and Monitoring of Non-Criteria (Toxic) Contaminants in Air*, Chicago, Illinois, March 1984, *Environ. Int.*, 1986, **12**, 221.

[25] P.W. Preuss, R.L. Dailey and E.S. Lehan, in *Formaldehyde: Analytical Chemistry and Toxicology*, ed. V. Turoski, American Chemical Society, Washington, 1985, p. 247.

[26] T.H. Stock and S.R. Mendez, *Am. Ind. Hyg. Assoc.*, 1985, **46**, 313.

[27] J.M. Samet, M.C. Marbury and J.D. Spengler, *Am. Rev. Respir. Dis.*, 1988, **137**, 221.

occur at the higher formaldehyde levels which arise due to UFFI or new furnishings and carpets. A range of effects reported among residents of US mobile homes[28] included burning eyes, cough, fatigue, dizziness, sore throat and wheeziness. The weekly average exposure was $11.9\,mg\,m^{-3}\,h^{-1}$. Some of the reported symptoms are known to be smoking related. A Canadian study[29] of respiratory symptoms, respiratory function and other effects in residents of UFFI-containing and control homes and a group of formaldehyde-exposed technicians found a higher prevalence of some non-specific symptoms in the UFFI home resident group compared with the two other groups, but no increase in prevalence of respiratory symptoms. The former finding is somewhat surprising, as the highest exposed group were the laboratory technicians, among whom there was no decrease in lung function. In a two-part study, also in Canada, comparing the health characteristics and respiratory function of occupants of homes containing UFFI and control homes both before and after remedial work or removal of UFFI, although some subjective measures of health appeared to be associated with formaldehyde or UFFI, there was no association between formaldehyde or UFFI and objective measures of lung function.[30-32] Some investigations have attempted to see if there are any specific effects among groups at extra risk from formaldehyde in the home. One study[33] demonstrated that asthma and bronchitis, but not other respiratory symptoms, were more prevalent among children in houses with higher (above $0.07\,mg\,m^{-3}$) formaldehyde concentrations; among adults, chronic cough in non-smokers was related to elevated formaldehyde levels, but respiratory symptoms and disease were not. Chamber studies showed no lung function changes when healthy volunteers were exposed to formaldehyde concentrations of 2.5 or $3.6\,mg\,m^{-3}$ for 40 or 180 minutes, respectively.[34,35] Similar studies, in which volunteers with a history of asthma and hyperactive airways were exposed to $3.6\,mg\,m^{-3}$ formaldehyde for 180 minutes, also demonstrated no effect on lung function.[36] In a further group of asthmatics exposed to 0, 0.12 or $0.85\,mg\,m^{-3}$ formaldehyde for approximately 90 minute periods in an exposure chamber, no exposure-related effects on lung function or bronchial reactivity were reported.[37]

Other Volatile Organic Compounds. Owing to differences in definitions of VOCs and TVOCs ('Total VOCs')[38] and in the efficiency with which individual

[28] K. S. Liu, F. Y. Huang, S. B. Hayward, J. Wesolowski and K. Sexton, *Environ. Health Perspect.*, 1991, **94**, 91.
[29] M. J. Bracken, D. J. Leasa and W. K. Morgan, *Can. J. Public Health*, 1985, **76**, 312.
[30] I. Broder, P. Corey, P. Cole, M. Lipa, S. Mintz and J. R. Nethercott, *Environ. Res.*, 1988, **45**, 141.
[31] I. Broder, P. Corey, P. Cole, M. Lipa, S. Mintz and J. R. Nethercott, *Environ. Res.*, 1988, **45**, 156.
[32] I. Broder, P. Corey, P. Brasher, M. Lipa and P. Cole, *Environ. Res.*, 1988, **45**, 179.
[33] M. Krzyanoski, J. J. Quackenboss and M. D. Lebowitz, *Environ. Res.*, 1990, **52**, 117.
[34] E. N. Schachter, T. J. Witek, T. Tosun and G. J. Beck, *Arch. Environ. Health*, 1986, **41**, 229.
[35] E. N. Schachter, T. J. Witek, D. J. Brody, T. Tosun, G. J. Beck and B. P. Leaderer, *Environ. Res.*, 1987, **44**, 188.
[36] L. R. Sauder, M. D. Chatham, D. J. Green and T. J. Kulle, *J. Occup. Med.*, 1986, **28**, 420.
[37] H. Harving, J. Korsgard, O. F. Pederson, L. Mølhave and R. Dahl, *Lung*, 1990, **168**, 15.
[38] ECA-IAQ, *Total Volatile Organic Compounds (TVOC) in Indoor Air Quality Investigations*, European Collaborative Action 'Indoor Air Quality and its Impact on Man', Report No 19. EUR 17675 EN, Office for Official Publications of the European Community, Luxembourg, 1995.

compounds, particularly very volatile organic compounds, are collected on absorbent materials, comparison of exposure data between studies is difficult. The most informative studies on exposure to VOCs in the home are the BRE study in the UK and the US EPA Total Personal Exposure Methodology (TEAM) Studies. Both of these large-scale investigations, although studying other exposures in the home and with different overall aims, included comprehensive elements to assess exposure to VOCs over a long period of time. In the BRE study,[39] over 200 individual VOCs were identified. The study found mean TVOC concentrations in different rooms were similar (0.2–0.4 mg m^{-3}) and indoor levels were ten times higher than those measured outdoors, which is broadly consistent with the US TEAM studies (see below). There was a significant relationship between increased TVOC concentrations and painting and decorating, the highest exposure to VOCs occurring during these activities. A number of guideline values for TVOCs from 5000 μg m^{-3} down to 200 μg m^{-3} are reported in the literature; the current BRE mean figure falls towards the lower end of this range. In the first part of the US TEAM study, conducted in New Jersey,[40] indoor levels of VOCs were found to be consistently higher than outdoor levels. There was a wide variation in individual exposure to specific compounds, breath levels of chloroform were related to levels in drinking water, and the strongest association was in breath analysis of benzene and styrene for smokers. For benzene, combining data from a number of elements of the TEAM study[41,42] it was estimated that 6 μg m^{-3} of an average personal exposure of 16 μg m^{-3} could be accounted for by outdoor air and the remaining 10 μg m^{-3} was due to personal activities (including smoking, which represented 50% of the exposure). Later analysis of the TEAM study[43] suggested that personal indoor air TVOC samples exceeded 1 mg m^{-3} in about 60% of all samples and 5 mg m^{-3} in about 10% of samples. A further study in North Carolina,[44] which was not part of the TEAM study, confirmed that peak exposures to VOCs were associated with painting and decorating activities and house cleaning.

Although large numbers of VOCs can be measured in indoor environments, most are present at levels that are orders of magnitude below those at which toxicological or even sensory effects would be expected in humans. However, they occur in variable and complex mixtures to which individuals are exposed for perhaps 80–90% of their time. Probably the most informative studies on health effects, albeit acute effects, of VOCs are gained from controlled chamber studies using defined concentrations of mixtures and defined endpoints. Studies by Otto

[39] V. M. Brown and D. R. Crump, in *Indoor Air Quality in Homes: Part 1, The Building Research Establishment Indoor Environment Study*, ed. R. W. Berry, V. M. Brown, D. R. Crump, M. Gavin, C. P. Grimes, D. F. Higham, A. V. Hull, C. A. Hunter, I. G. Jeffery, R. G. Lea, J. W. Llewellyn and G. J. Raw, Construction Research Communications, London, 1996, p. 38.

[40] L. A. Wallace, E. D. Pellizzari, T. D. Hartwell, R. Whitmore, C. Sparino and H. Zelon, *Environ. Int.*, 1986, **12**, 369.

[41] L. A. Wallace, *J. Am. Coll. Toxicol.*, 1989, **8**, 883.

[42] L. A. Wallace, *Risk Anal.*, 1990, **10**, 59.

[43] L. A. Wallace, *Ann. N. Y. Acad. Sci.*, 1992, **641**, 7.

[44] L. A. Wallace, E. D. Pellizzari, T. D. Hartwell, V. Davis, L. C. Michael and R. W. Whitmore, *Environ. Res.*, 1989, **50**, 37.

et al.,[45] using a mixture of 22 VOCs at $25\,mg\,m^{-3}$ (total), evoked irritancy symptoms, as measured by questionnaire response, but no neurobehavioural changes. Other studies with VOC mixtures, also in chamber situations, have evoked positive subjective responses to air quality in a group reporting 'sick building syndrome' (SBS) symptoms at work, but not in a control group,[46] and possible lung function changes among a group of non-smoking volunteers.[47] However, in neither study was a clear association between VOC exposure and effect found. VOCs may also have a role in the perception of air quality, but this is not easy to separate from other factors contributing to the overall odour of indoor air. Apart from odour recognition itself, the perception of unpleasant odour may signify poor air quality and lead to or trigger secondary effects.[48] Various methods, including the use of trained panels to make subjective evaluations of perceived air quality in relation to occupant comfort using quantitative descriptions of pollution emissions and air quality, have been developed over the last few years,[49] although the methods are not universally accepted. Moreover, there appears to be no consistent association between occupant dissatisfaction with air quality and odour perceptions, or between perceived air quality and TVOC level. Factors such as temperature and humidity have also been reported to be important determinants of perceived air quality and of SBS symptoms[50,51] (see below). The subject of sensory perception of air quality is under current review by the European Concerted Action on Indoor Air Quality and its Impact on Man.

Conclusions. With regard to formaldehyde, chamber studies with normal adults or those with pre-existing asthma have not demonstrated any dysfunction at mean formaldehyde levels typically found in homes, or even at levels several times higher. Moreover, epidemiological studies have not demonstrated any increase in respiratory symptoms or lung function at estimated current domestic levels. It is concluded that most people would fail to show any sensory effects at an ambient maximum concentration of $0.1\,mg\,m^{-3}$ averaged over $0.5\,h$, although some individuals might show transient effects at or below this level. For the protection of health, exposure to formaldehyde in the domestic environment should remain at or below current levels; significantly higher levels should be avoided. There are a number of research needs on the health effects of domestic formaldehyde exposure. For example, the incidence and nature of hyper-reactivity to formaldehyde should be studied across a wide range of concentrations, and the effects of combined exposures to formaldehyde and other common household substances

---

[45] D. Otto, L. Mølhave, G. Rose, H. K. Hudnell and D. House, *Neurotoxicol. Teratol.*, 1990, **12**, 649.

[46] S. K. Kjærgaard, L. Mølhave and O. F. Pederson, *Atmos. Environ.*, 1991, **25A**, 1417.

[47] H. Harving, R. Dahl and L. Mølhave, *Am. Rev. Respir. Dis.*, 1991, **143**, 751.

[48] WHO, *Indoor Air Quality: Organic Pollutants*, Euro Reports and Studies No. 111, World Health Organization, Copenhagen, 1989.

[49] P. O. Fanger, *Energy Build.*, 1988, **12**, 1.

[50] L. Mølhave, S. K. Kjærgaard, O. F. Pederson, A. H. Jorgenson and T. Pedersen, in *Proceedings of Indoor Air '93*, Helsinki, 1993, p. 555.

[51] L. Berglund and W. S. Cain, in *Proceedings of Indoor Air Quality '89: The Human Health Equation*, ASHRAE, Atlanta, 1989, p. 93.

should be investigated. There are many formaldehyde sources in the home, so any control strategy has to take account of this multiplicity of sources.

Regarding other VOCs, there is no evidence to suggest that current typical indoor (domestic) exposure to VOCs—either individually or as a total—poses a health risk. Based on chamber studies, TVOCs at concentrations greater than $25\,mg\,m^{-3}$ may cause acute irritancy and other transient effects; although such concentrations are unlikely to be encountered under normal domestic conditions, they could occur during painting/decorating or excessive solvent usage. The composition of TVOCs is complex and variable and health effects resulting from exposure are generally poorly characterized. It is, therefore, prudent to minimize exposure to TVOCs, particularly genotoxic and carcinogenic substances. For certain specific VOCs, such as benzene, appropriate guidelines may be applied. In the case of benzene, the UK Expert Panel on Air Quality Standards (EPAQS) recommended standard of $16.2\,\mu g\,m^{-3}$ (5 ppb) running average, together with the recommendation to reduce overall levels of exposure to benzene such that (for non-smokers) ambient air pollution is no longer the main source of individual exposure,[52] are to be encouraged. Information should be available to people about the most important sources of VOCs in the home, including activities that lead to exposure, so that they may consider how to minimize their exposure and any associated effects. Also, consideration should be given to monitoring indoor air quality in homes in order to assess the effectiveness of control measures applied, for example, to building materials and techniques, and to consumer products. With respect to outstanding research needs on VOCs, further toxicological data should be collected on individual VOCs, and on their sensory thresholds, and methods should be improved for the evaluation of the sensory and neuropsychological effects of VOCs.

Fungi and Bacteria

Damp and mould are relatively common conditions in European housing, and there is a history of concern regarding the possible effects on health of exposure to fungal spores (and other fungal-derived material) and, to a lesser extent, bacteria. Many different species of bacteria and fungi can be found in homes, associated with various forms of organic matter such as surface coating of walls, wood, fabrics and foodstuffs. Some species are particularly associated with dampness in buildings and several health effects (other than infections) have been attributed to the saprotrophic bacterial and fungal flora of the indoor environment.

In the recent BRE study,[53] two different methods were used to sample airborne fungi and bacteria: a filter technique and a multi-stage Andersen sampler. The data from these two sampling methods were comparable, both in total numbers and species present, to other studies reported from the UK and elsewhere. The

[52] Expert Panel on Air Quality Standards, *Benzene*, HMSO, London, 1994.

[53] C.A. Hunter, A.V. Hull, D.F. Higham, C.P. Grimes and R.G. Lea, in *Indoor Air Quality in Homes: Part 1, The Building Research Establishment Indoor Environment Study*, ed. R.W. Berry, S.K.D. Coward, D.R. Crump, M. Gavin, C.P. Grimes, D.F. Higham, A.V. Hull, C.A. Hunter, I.G. Jeffrey, R.G. Lea, J.W. Llewellyn and G.J. Raw, Construction Research Communications, London, 1996, p. 99.

isolates identified were not atypical, with *Penicillium, Cladosporium, Aspergillus* and *Mycelia sterilia* predominating. It should be noted, however, that minor species, which could pose a biological hazard, might not be detected and the amount of cultivable organisms may only represent a small and variable fraction of the total airborne flora. Longer sampling periods associated with the filter method might give a more representative estimate of levels than the shorter sampling period used with the Andersen sampler, but like all other sampling methods, there are certain intractable technical deficiencies. As with fungi, the concentrations of bacteria obtained with the two sampling methods were of the same order of magnitude as found in other studies. In addition, Gram-positive bacteria predominated over Gram-negative types, which is compatible with previous findings. However, the rank order of the bacterial genera differed slightly between the two methods and this may be due to the desiccation effect on the non-spore-forming bacteria seen with the filter collection method. From other studies, it can be concluded that the numbers of viable bacteria recorded would form an even smaller proportion of the total count than is the case for fungi.

A number of epidemiological studies conducted in Europe and North America have investigated the relation between home dampness and respiratory morbidity in children and adults. In most of these studies, information about the exposure variables (home dampness and mould) and the outcome variables (respiratory symptoms) was obtained by questionnaires. Only in a few of the studies was further information about exposure to moulds obtained by actually measuring the number of airborne propagules. Two such studies reported a positive association between airborne fungal counts and some respiratory symptoms in children,[54,55] whereas a third found no such association.[56] A fourth study found no association between fungal counts in house dust samples and respiratory symptoms in children.[57] The limited number of studies linking the measured levels of airborne organisms in the home with adverse health outcomes all relate to fungi; there appear to be none concerning bacteria.[58]

Conclusions. There is consistent evidence of an association between damp and mouldy housing and reports of respiratory symptoms in children. However, the causal interpretation of these findings remains uncertain. Numerous fungal and bacterial species are present in homes and the health effects of many species and their products are unknown or poorly understood. Epidemiological studies relating measurements of indoor airborne fungi to respiratory disease generally have not shown convincing associations. It is not clear whether this is due to the recognized limitations of current mycological methods in providing an index of relevant exposure, or to the true absence of a health effect. The literature relating domestic mould growth to non-respiratory disease is extremely sparse and

[54] S. D. Platt, C. J. Martin, S. M. Hunt and C. W. Lewis, *Br. Med. J.*, 1989, **298**, 1673.
[55] M. Waegemaekers, N. Van Wageningen, B. Brunekreef and J. S. M. Boliej, *Allergy*, 1989, **44**, 192.
[56] D. P. Strachan, B. Flannigan, E. M. McCabe and F. McGarry, *Thorax*, 1990, **45**, 382.
[57] A. P. Verhoeff, J. H. Van Wijnen, E. S. Van Reene-Hockstra, R. A. Samson, R. T. Van Strien and B. Brunekreef, *Allergy*, 1994, **49**, 540.
[58] B. Flannigan, in *Clean Air at Work*, ed. R. H. Brown, M. Curtis, K. J. Saunders and S. Vandendriessche, The Royal Society of Chemistry, Cambridge, 1992, p. 366.

although there are recognized health hazards, no epidemiological data exist to quantitate exposure–response relationships. A number of studies have drawn attention to a relationship between dampness and mould growth in houses and symptoms of respiratory disease in their occupants, but these relationships cannot at present be attributed to specific fungi or bacteria in the air. Mould and dampness are often associated with poor ventilation, which tends to increase exposure to other contaminants as well as microbiological products. Improved ventilation could be expected to reduce indoor dampness and mould growth.

There appear to be complicated inter-relationships between dampness and other building factors, heating (type and degree), the presence of mould and socio-economic factors in the association with occupants' ill-health. There is a dearth of information on the toxicity of fungi and bacteria and their metabolites, which needs to be addressed, and the general issue of damp housing and health similarly requires further study.

House Dust Mites

It is well established that house dust mites are ubiquitous in homes in warm temperature regions and that their relative abundance is largely determined by the internal microclimate, since they tend to prefer warm damp conditions. There is clear evidence that antigen derived largely from house mite faeces is one of the major causes of allergic sensitization and that people who have been sensitized to mites are more likely than those not sensitized to manifest symptoms of asthma and other allergies.

Temperature and humidity are important factors affecting the distribution and abundance of house dust mites, influencing the quantity of mite allergens, the species of mites found and the distribution of mites within a house. Increased ventilation and air conditioning is associated with lower levels of mite allergens and has been shown to reduce seasonal increases of mite allergens in the US.[59] A recent study by the BRE,[60] looking at homes in the county of Avon, UK, found high numbers of mites in both living room and bedroom carpets; 95% of mites sampled were *Dermatophagoides pteronyssinus* (from which the allergen Der p1 is derived). A correlation between relative humidity and mite numbers was confirmed. Mite counts typically used in earlier studies are technically demanding, time consuming and may underestimate the number of live mites.[61] The problems with the methodology relate to several stages in the analysis. Sampling strategy has not been standardized and the actual methodology seems to vary from study to study. Extraction of mites from dust is not necessarily quantitative. Nevertheless, this method does allow species identification and can be useful in intervention studies, as decline in mite numbers may precede changes in antigen levels.[62] More recently, with the development of sensitive and specific monoclonal

[59] C. M. Luczynska, *Respir. Med.*, 1994, **88**, 723.

[60] C. A. Hunter, I. G. Jeffrey, R. W. Berry and R. G. Lea, in *Indoor Air Quality in Homes: Part 1, The Building Research Establishment Indoor Environment Study*, ed. R. W. Berry, S. K. D. Coward, D. R. Crump, M. Gavin, C. P. Grimes, D. F. Higham, A. V. Hull, C. A. Hunter, I. G. Jeffrey, R. G. Lea, J. W. Llewellyn and G. J. Raw, Construction Research Communications, London, 1996, p. 87.

[61] T. A. E. Platts-Mills and A. L. de Weck, *J. Allergy Clin. Immunol.*, 1989, **83**, 416.

immunoassays, it has been possible to quantify concentrations of mite antigens in dust,[63] but again there are problems of sampling strategy and technique. Occasional studies have used guanine as an indirect measure of allergen in dust.[64] This is a semi-quantitative technique that poorly correlates with allergen levels, but can be useful as a screening tool and has been used to monitor interventions. Measurements of allergen in air are critically dependent on sampling strategy. Domestic activity and the particle size associated with the allergen affect both the quantities of airborne allergen and the duration that allergens are airborne.[65] There is clear evidence from the studies that have been performed that antigen derived largely from mite faeces is one of the major causes of allergic sensitization.

Sensitization is much more likely to occur in people who are predisposed to the development of atopic disease on the basis of genetic predisposition and other as yet unknown environmental factors. People who have been sensitized to mites are more likely than those not sensitized to manifest symptoms of asthma and other allergies. Moreover, those sensitized to mites are likely to develop symptoms in response to exposure to the mites, either in the natural circumstances of house dust exposure or in the artificial circumstances of bronchial challenge. Furthermore, it has been shown in studies intended to reduce exposure of symptomatic individuals to mite antigen that a reduction in symptoms may occur. There is thus evidence that exposure to mites is a hazard with respect to development of sensitization, initiation of asthma and provocation of asthmatic symptoms.[61,62] However, it is far from clear whether mites are responsible, in whole or in part, for the general rise in the prevalence of asthma in the UK and elsewhere. There is no convincing evidence that there has been sufficient change in mite populations in houses to explain such a change,[66] and there are reasons to suppose that other factors relating to susceptibility are likely to be of additional importance. This is relevant, as a major effort to reduce mite populations in houses may not influence substantially the prevalence of asthma and allergic disease in the population, since susceptible people would still become sensitized to other common allergens. It is possible, however, that such measures would reduce the severity of symptoms in people already sensitized to mites.[67] Comparisons of disease prevalence in populations with different levels of mite allergen exposure are prone to confounding by other environmental exposures, or by differences in genetic constitution or lifestyle. For these reasons they are difficult to interpret as evidence either for or against an effect of mite allergen exposure on asthma prevalence, particularly when they are based on a comparison of only two study centres or populations. Similarly, changes in asthma prevalence over time may be due to factors other than changes in mite

[62] T. A. E. Platts-Mills, W. Thomas, R. C. Aalberse, D. Vervoet and M. D. Chapman, *J. Allergy Clin. Immunol.*, 1992, **89**, 1046.

[63] M. D. Chapman, P. W. Heymann, S. R. Wilkins, M. J. Brown and T. A. E. Platts-Mills, *J. Allergy Clin. Immunol.*, 1987, **80**, 184.

[64] J. E. M. H. Van Bronswijk, E. Bischoff, W. Schmiracher and F. M. Kniest, *J. Med. Entomol.*, 1989, **26**, 55.

[65] F. De Blay, P. W. Heymann, M. D. Chapman and T. A. E. Platts-Mills, *J. Allergy Clin. Immunol.*, 1991, **88**, 919.

[66] R. Sporik, S. T. Holgate, T. A. E. Platts-Mills and J. J. Cogswell, *New. Engl. J. Med.*, 1990, **323**, 502.

[67] M. J. Colloff, *Br. J. Dermatol.*, 1992, **127**, 322.

allergen exposure. A striking and widely quoted epidemic of asthma in the Fore region of Papua New Guinea[68] was attributed to the introduction of dust mite in blankets, but there were profound changes in many other aspects of the highlanders' lifestyle at the same time. Changes in mite allergen exposure at much higher levels do not appear to have contributed to recent increases in asthma prevalence (*e.g.*, in Australia[69] or the UK[66]). Most studies which have investigated the relationship between mite allergen exposure and mite sensitization within a population relate to children. A positive correlation has been reported from various centres with differing levels of allergen exposure, although the imprecision of a single cross-sectional exposure measurement leaves open the possibility that children apparently sensitized at very low levels of current exposure in such studies may have been exposed to high levels in the past. Thus it is not possible to determine whether there is a threshold exposure level below which sensitization does not occur. The importance of genetic predisposition in defining the position and shape of the exposure–sensitization relationship is recognized.[70,71] It is likely that there are some people who would not become sensitized even at very high levels of mite allergen exposure, but a plateau in the exposure–sensitization relationship at high levels has yet to be demonstrated by epidemiological studies.

For respiratory disease to develop, a series of steps must occur in a genetically predisposed individual. These are sensitization, the development of bronchial reactivity, and finally a response to continued exposure, producing symptoms or changes in lung function. A number of studies have attempted to assess the exposure–response relationship or threshold at which symptoms will occur when a sensitized individual is exposed. Most of these studies have been either birth cohort studies of asthma incidence, or cross-sectional studies of disease prevalence or severity, mainly in children. Two prospective studies of infants at higher risk of allergy have failed to offer conclusive evidence of a positive relationship. The widely cited paper by Sporik *et al.*[66] suggested that at higher levels of exposure in infancy (greater than $10\,\mu g$ Der p1 g^{-1} dust), there is an increase in asthmatic symptoms up to age 11. However, this finding was of borderline statistical significance and the more convincing relationship was between early allergen exposure and an early age of onset of wheezing. A similarly designed study showed no association between allergen exposure in infancy and mite sensitization at age seven years.[72] The mite allergen concentration in the first and seventh years of life did not differ significantly between atopic children with and without a history of wheezing. A third cohort study of infants at high risk of allergy, which involved intervention, was also not suggestive of an

[68] G.K. Dowse, K.J. Turner, G.A. Stewart, M.P. Alpers and A.J. Woolcock, *J. Allergy Clin. Immunol.*, 1985, **75**, 75.
[69] J.K. Peat, R.H. Van den Berg, W.F. Green, C.M. Mellis and S.R. Leeder, *Br. Med. J.*, 1994, **308**, 1591.
[70] J. Kuehr, T. Frischer, R. Meinert, R. Barth, J. Forster, S. Schraub, R. Urbanek and W. Karmaus, *J. Allergy Clin. Immunol.*, 1994, **94**, 44.
[71] R.P. Young, B.J. Hart, T.G. Merrett, A.F. Read and J.M. Hopkin, *Clin. Exp. Allergy*, 1992, **22**, 205.
[72] M.L. Burr, E.S. Limb, M.J. Maguire, L. Amarah, B.A. Eldridge, J.C.M. Layzell and T.G. Merrett, *Arch. Dis. Child.*, 1993, **68**, 724.

exposure– response relationship for symptoms.[73] These findings are consistent with various cross-sectional studies[74,75] in which no significant or substantial association has emerged between mite allergen exposure and the prevalence of asthma symptoms in children, at exposures generally higher than $2\,\mu\mathrm{g}\,\mathrm{g}^{-1}$ dust.

Conclusions. Within the range of allergen exposures commonly encountered in homes (greater than $2\,\mu\mathrm{g}\,\mathrm{g}^{-1}$ dust), there is fairly consistent evidence of an increase in risk of mite sensitization with increasing allergen exposure. However, there is also evidence that allergen exposure may influence the risk of sensitization below the $2\,\mu\mathrm{g}\,\mathrm{g}^{-1}$ threshold. At all detectable levels of mite allergen exposure, a reduction may be expected to reduce the risk of mite sensitization.

Mite sensitization does not inevitably result in mite-sensitive asthma. The shape of the exposure–response relationship relating asthmatic symptoms (such as wheeze) to allergen exposure among sensitized subjects may be different from that relating allergen exposure to sensitization. The evidence from both cross-sectional and longitudinal studies is consistent with a saturation or plateau effect at levels of allergen exposure currently encountered in many homes. This implies that there may be little change in prevalence of asthma associated with a modest downward shift in allergen levels. The observational evidence relating prior allergen exposure to acute asthma attacks is inconsistent. Evidence from a large number of small controlled trials of diverse allergen reduction regimens suggests that there is little clinical benefit unless allergen levels are reduced substantially. However, each individual study lacks statistical power to demonstrate a small benefit which would nevertheless be of significance in public health terms. A formal meta-analysis is not possible because of the diversity of the outcome measures and exposure assessments in the different trials. Thus it is possible that reduction in the allergen exposure of asthmatic patients might result in a small reduction in morbidity, but the extent of the health gain (if any) cannot be quantified.

In conclusion, there is no convincing evidence of a strong exposure–response relationship between asthma symptoms and house dust mite allergen at the levels of exposure presently encountered. It may be that exposures are at the plateau of a non-linear exposure–response curve and would need to be significantly lower before a strong relationship is identified. Further studies are required to assess whether changes in the indoor environment, with and without changes in lifestyle, may lead to reduced exposure and decreases in symptomatic asthma. Further investigations are needed of the effectiveness of allergen reduction regimes, in terms of impact on allergen exposure and on incidence, prevalence and severity of symptoms. Studies should also be undertaken to clarify the exposure–response relationship between allergic sensitization, symptom prevalence and disease severity and exposure to house dust mites or mite allergens. Despite the uncertainties about the exposure–response relationship(s), a general reduction in mite allergen exposure in homes is encouraged. Lower indoor humidity could contribute to reducing mite numbers and therefore exposure to allergen.

[73] D. W. Hide, S. Matthews, L. Matthews, M. Stevens, S. Ridout, R. Twiselton, C. Gant and S. H. Arshad, *J. Allergy Clin. Immunol.*, 1994, **93**, 842.

[74] K. M. Strachan, B. K. Paine, B. K. Butland and H. R. Anderson, *Thorax*, 1993, **48**, 426.

[75] A. P. Verhoeff, R. T. Van Strien, J. H. Van Wijen and B. Brunekreef, *Clin. Exp. Allergy*, 1994, **24**, 1061.

Carbon Monoxide (CO)

Carbon monoxide is of particular interest and importance as many deaths and hospital admissions can be directly attributable to accidental domestic CO poisoning.[76] It is especially dangerous because it has no colour, smell or taste. Its toxic action is mostly through the displacement of oxygen in haemoglobin in the blood to form carboxyhaemoglobin, thus depriving the tissues of the body of their oxygen supply. Most fatal cases of carbon monoxide poisoning result from blockage of and/or leakage from flues of gas heating appliances.

There is a large body of literature concerning indoor concentrations and the health effects of CO (although very few studies have to date been conducted in the UK). Outdoor CO levels can be determinants of indoor levels but, where present, the major sources of CO in the home are gas cookers and certain types of heating systems that burn gas, wood, coal or paraffin. Environmental tobacco smoke, the presence of an attached garage and the proximity of heavily trafficked roads can also affect indoor CO levels.[77]

A recent UK study has shown typical 1-week average CO concentrations to reach $2.7 \, \text{mg m}^{-3}$ (2.4 ppm) in the kitchens of homes where there was gas cooking, compared with $0.9 \, \text{mg m}^{-3}$ (0.79 ppm) in kitchens where there was no gas cooking. Continuous monitoring indicated maximum 1-hour averages of $1.9 – 24.5 \, \text{mg m}^{-3}$ (1.7–21.4 ppm) in homes with gas cooking;[78] much higher peak levels of around $180 \, \text{mg m}^{-3}$ (160 ppm) for a 15-minute average have been associated with the use of a gas cooker grill.[79] Poorly installed, inadequately ventilated or malfunctioning appliances and accidentally blocked flues can also contribute to increased CO levels. Even in a sample of only 14 UK homes, a maximum 1-hour concentration of $57.0 \, \text{mg m}^{-3}$ (50 ppm) was recorded in the kitchen of one home in which the boiler was malfunctioning.[78] It is apparent that existing air quality guidelines[80]* are likely to be exceeded in a number of homes. While it is not statistically valid to extrapolate the data from the small study of 14 homes in the UK to the overall national (or international) situation, there is an obvious cause for concern.

Exposure to CO is normally evaluated in terms of percentage of carboxyhaemoglobin (COHb) in the blood, but the validity of COHb as a biomarker of health effect is open to question. Smokers have higher blood COHb levels and a higher threshold of effects. Although hypoxia, arising from preferential binding of CO to haemoglobin, is thought to be the main toxic mechanism by which CO acts, binding of CO to other blood components and enzymes may also play a part

---

* The current World Health Organization guidelines for CO are $100 \, \text{mg m}^{-3}$ for 15 minutes, $60 \, \text{mg m}^{-3}$ for 30 minutes, $30 \, \text{mg m}^{-3}$ for 1 hour and $10 \, \text{mg m}^{-3}$ for 8 hours.

[76] M. Burr, in *Building Regulation and Health*, ed. G.J. Raw and R.M. Hamilton, Construction Research Communications, London, 1995, p. 26.

[77] IEH, *Assessment on Indoor Air Quality in the Home 2: Carbon Monoxide*, Assessment A5, Institute for Environment and Health, Leicester, 1998, in press.

[78] D. Ross, in *Proceedings of the 7th International Conference on Indoor Air Quality and Climate*, ed. S. Yoshizawa, K.-i. Kimura, I. Ikeda, S.-i. Tanabe and T. Iwata, Institute for Public Health, Tokyo, 1996, p. 513.

[79] K.J. Stevenson, *Tokai J. Exp. Clin. Med.*, 1985, **10**, 295.

[80] WHO, *Update and Revision of the Air Quality Guidelines for Europe*, World Health Organization Regional Office for Europe, Copenhagen, 1994.

in its toxicity. A role in promoting atherosclerosis has been postulated for CO, although conclusive evidence is lacking, and immunological function and neurotransmission have also been investigated as possible targets for CO toxicity.[81]

Carbon monoxide is an important pollutant with respect to likely health effects following exposure in the home. While many of the published clinical investigations of CO intoxication in the home originate outside the UK, this does not limit their applicability. There may be differences in the types of cooking and heating appliances used but the health effects of the CO emitted from them will be broadly the same. Accidental exposures leading to acute, and sometimes fatal, health effects are well documented. Clinical reports of CO intoxication following exposure to high levels of CO have shown consistent symptoms such as headache, nausea and dizziness in the majority of patients. However, these symptoms are easily confused with those of other ailments, such as food poisoning or influenza, and missed or mis-diagnoses of CO intoxication can therefore occur.[82]

Numerous and varied observations have been made of the health effects of CO in controlled chamber studies.[81] These indicate that exposure to CO can cause performance decrements in certain neuropsychological tasks and that some people, primarily suffers of cardiovascular disease, may be more susceptible to low level exposure to CO associated with COHb levels as low as 2%.[83] However, the question of the COHb level at which cardiovascular indices do not differ from the norm has not been satisfactorily answered.[77]

Conclusions. The published evidence on health effects after domestic exposure points most to a hazard of acute CO intoxication from malfunctioning, unflued or poorly ventilated fuel burning appliances. It is also probable that in some homes CO levels routinely occur and persist that might possibly give rise to chronic health effects, particularly among sensitive groups (pregnant mothers, the foetus, children, the elderly and individuals suffering from anaemia and other diseases that restrict oxygen transport). Significant symptoms are generally experienced, even among normal healthy individuals, following exposure to CO concentrations high enough to produce COHb levels of about 20%. A great deal of importance would be attached to CO concentrations producing COHb levels above 10%, especially in sensitive individuals.

Although there is limited information from epidemiological studies on the health effects of CO at the low levels typically found in homes, the risk of adverse effects in healthy individuals as a result of exposure to CO in the home is thought to be low under normal circumstances (*i.e.* where appliances are installed and operated correctly). Nonetheless, it is prudent to continue to encourage measures which minimize CO levels, with particular attention being paid to gas combustion and other fuel-burning, especially unflued, appliances. It is also essential to increase awareness of the symptomatology of CO intoxication among health care professionals and others to whom the public look for advice and assistance.

[81] EPA, *Air Quality Criteria for Carbon Monoxide*, EPA/600/8-90/045F, US Environmental Protection Agency, Cincinnati, 1991.

[82] F. L. Lowe-Ponsford and J. A. Henry, *Advers. Drug React. Acute Poison. Rev.*, 1989, **8**, 217.

[83] E. N. Allred, E. R. Bleecker, B. R. Chaitman, T. E. Dahms, S. O. Gottlieb, J. D. Hackney, M. Pagano, R. H. Selvester, S. M. Walden and J. Waren, *Environ. Health Perspect.*, 1991, **91**, 89.

Leaving a patient in, or returning them to, a situation from which adverse health effects might develop is unacceptable and, with vigilance, need not occur.

There are a number of areas requiring further research.[77] In particular, more studies are required to determine the importance of indoor levels to overall personal exposure to CO, especially the significance of certain activities and situations which may lead to high exposures. It would also be of value to investigate exposure to CO in susceptible populations such as expectant mothers and those suffering from cardiovascular disease.

Particles (PM_{10})*

Particulates continue to attract a good deal of attention as a possible major cause of early deaths in the population, as revealed for example by the US Six Cities study investigating the health impacts of outdoor PM_{10} levels.[84]

The major indoor sources of particles have been identified through studies performed in the US (*e.g.* the Harvard Six-City Study, the New York State Study and the EPA Particle Total Personal Exposure Assessment (PTEAM) Study).[85-88] Environmental tobacco smoke (ETS, considered separately below) has consistently been shown to be the most significant indoor particle source. Emissions from kerosene heaters and wood-burning stoves have also been shown to add to the indoor particle load, although these sources are of less significance in the UK and other countries. Some US studies have also identified cooking as a source of particles (from both the food itself and combustion of the cooking fuel).[87,89] Very fine particles are thought to readily enter buildings[90] so outdoor sources (*e.g.* traffic) are also important to personal exposure. The activity of occupants in the home has also been found to influence indoor particle levels. Vacuuming, sweeping and dusting have been shown to raise levels of particles.[87,89] Although these are not direct sources of particles, in that they represent re-entrainment of settled particles, such activities may affect the total indoor level to which occupants are exposed. Indeed, a 'personal cloud' effect has been described whereby personal exposure exceeds that expected from statically monitored indoor and outdoor levels.[91] This suggests that human behaviour and activity can markedly influence exposure.

Limited information suggests that indoor levels of particles are generally lower

* Particulate matter of 10 μm or less aerodynamic diameter.

[84] D. W. Dockery, C. A. Pope, III, X. Xu, J. D. Spengler, H. H. Ware, M. E. Fay, B. G. Ferris, Jr. and F. E. Speizer, *New Engl. J. Med.*, 1993, **329**, 1753.

[85] J. D. Spengler, D. W. Dockery, W. A. Turner, J. M. Wolfson and B. G. Ferris, *Atmos. Environ.*, 1981, **15**, 23.

[86] J. D. Spengler, R. D. Treitman, T. D. Toteson, D. T. Mage and M. L. Soczek, *Environ. Sci. Technol.*, 1985, **19**, 700.

[87] H. Özkaynak, J. Xue, J. Spengler, L. Wallace, E. Pellizzari and P. Jenkins, *J. Expos. Anal. Environ. Epidemiol.*, 1996, **6**, 57.

[88] B. P. Leaderer, P. Koutrakis, S. L. K. Briggs and J. Rizzuto, *Indoor Air Int. J. Indoor Air Qual. Clim.*, 1994, **4**, 23.

[89] R. Kamens, C.-T. Lee, R. Wiener and D. Leith, *Atmos. Environ., Part A*, 1991, **25**, 939.

[90] T. L. Thatcher and D. W. Layton, *Atmos. Environ.*, 1995, **29**, 1487.

[91] C. A. Clayton, R. L. Perritt, E. D. Pellizzari, K. W. Thomas, R. W. Whitmore, L. A. Wallace, H. Özkaynak and J. D. Spengler, *J. Expos. Anal. Environ. Epidemiol.*, 1993, **3**, 227.

than, but correlated to, outdoor levels unless there is a significant indoor source. Thus where there are no major indoor sources of particles, outdoor levels may be a reasonable proxy for indoor exposure, but they cannot accurately estimate *personal exposure* to particulates because of the 'personal cloud' effect.

Epidemiological studies have demonstrated a consistent and statistically significant association between ambient (outdoor) airborne particle level and the incidence of mortality or morbidity in human populations. The associations with death and hospital/emergency room admission rates are the most robust[92,93] and appear to operate at low exposure levels within the range frequently encountered in many developed countries.[94] Cardiopulmonary impairment appears to be the predominant effect, and the elderly or infirm appear to be at particular risk. In addition, irrespective of age, asthmatics appear to suffer increased symptomatology and increased risk of acute attack.[92,95,96] There is also evidence that ambient particle exposure is associated with falls in pulmonary function measures, especially for asthmatics.[97,98] Chronic effects on cardiopulmonary disease and, possibly, cancer have been noted,[84] while there are reports from China of small reductions in the duration of pregnancy and in birth weight.[99] There is also some evidence from the US for effects on postnatal mortality.[100] Despite the consistency of some of these findings, interpretation and comparison is difficult, not only because of the implicit limitations of epidemiological studies, but also because of the widely differing approaches taken to the classification and monitoring of the particulate fraction of ambient air. Because of this, the validity of the available epidemiology data on outdoor particle exposure has been questioned by some workers, and there is a need for better quantitative risk estimates of the long-term impact of PM_{10} exposure.[96,101] There is a dearth of studies investigating possible links between health effects and indoor or personal exposure to particles or other airborne pollutants such that, at present, these aspects cannot be confidently assessed.

In addition to the epidemiological evidence, there is some support for a causal link between non-ETS-derived particles and adverse health effects from human volunteer studies. In summarizing the status of knowledge, the UK Committee on the Medical Effects of Air Pollutants[95] noted that there had been few investigations, mostly focusing on sulfuric or other acid aerosols. There is evidence for effects on lung function and bronchial reactivity, although the effects

[92] D. W. Dockery and C. A. Pope, III, *Annu. Rev. Public Health*, 1994, **15**, 107.

[93] K. Katsouyanni, G. Touloumi, C. Spix, J. Schwartz, F. Balducci, S. Medina, G. Rossi, B. Wojtyniak, J. Sunyer, L. Bacharova, J. P. Schouten, A. Pönkä and H. R. Anderson, *Br. Med. J.*, 1997, **314**, 1658.

[94] B. Brunekreef, D. W. Dockery and M. Krzyzanowski, *Environ. Health Perspect.*, 1995, **103** (suppl. 2), 3.

[95] COMEAP, *Non-biological Particles and Health*, Committee on the Medical Effects of Air Pollution, HMSO, London, 1995.

[96] C. A. Pope, III, D. W. Dockery and J. Schwartz, *Inhal. Toxicol.*, 1995, **7**, 1.

[97] J. Q. Koenig, K. Dumler, V. Rebolledo, P. V. Williams and W. E. Pierson, *Arch. Environ. Health*, 1993, **48**, 171.

[98] L. M. Neas, D. W. Dockery, P. Koutrakis, D. J. Tollerud and F. E. Speizer, *Am. J. Epidemiol.*, 1995, **141**, 111.

[99] X. Xu, H. Ding and X. Wang, *Arch. Environ. Health*, 1995, **50**, 407.

[100] M. Bobak and D. A. Leon, *Lancet*, 1992, **340**, 1010.

[101] C. A. Pope, III, D. V. Bates and M. E. Raizenne, *Environ. Health Perspect.*, 1995, **103**, 472.

reported have been variable. In normal subjects, slight changes in lung function were noted at high concentrations ($100 \, \mu g \, m^{-3}$) of sulfuric acid aerosol, while exposure to diesel exhaust at a concentration of $4.3 \times 10^6 \, cm^{-3}$ was shown to elicit pulmonary inflammatory changes.[102,103] Overall, although human studies have confirmed that some particles can cause physiological change in healthy humans, this has only been demonstrated at exposures above those normally experienced in the environment. However, human volunteer studies have found differences between population sub-groups, for example regional deposition patterns within the respiratory tract differ between children and adults,[104] and this will result in different doses and sites of deposition at identical exposures. Also, as noted above, asthmatics seem to be more sensitive to the effects of particles. Non-human experimental work has confirmed the intrinsic toxicity of some particles and has suggested a number of possible toxic mechanisms. However, the relative importance of particle number, size, mass and composition is still not clear.

Other hypotheses (not assuming a causal relationship) have been put forward to explain the positive correlation between outdoor particle levels and mortality or morbidity rates observed in epidemiological studies. Valberg and Watson,[105] for example, support the view that there is no causal link between excess mortality or morbidity and exposure to particulate matter, but that particle levels are merely a marker for other causal factors (including increased exposure to indoor pollutants).

The role played by sources which are not of outdoor origin in determining personal exposure to particles needs to be elucidated. A key question concerns the contribution of indoor particulate sources to total personal exposure to particles, and hence the potential impact of indoor particle sources on health. If indoor particles significantly influence personal exposure levels, then the existing epidemiological evidence, which is based upon outdoor levels of particles, would need to be revisited and reinterpreted.

Conclusions. Epidemiological studies have shown that those at greatest risk from exposure to particles are people with existing respiratory or cardiovascular diseases. Many of these people may be elderly or infirm, and would be expected to spend the majority of their time indoors. Therefore, it may be reasonable to assume that personal exposure to particles in this group is most influenced by the level of particles in indoor air. Little is known about the level or composition of particles to which such susceptible groups are exposed, and whether this differs from the exposure of the general population. Whilst a link between particulates in outdoor air and ill health has been demonstrated through epidemiological studies, there have been no similar studies based on particle levels in indoor air.

Assuming that the relationship between particle exposure and effects on health is causal, remedial strategies should be aimed at reducing total personal exposure

[102] M. T. Newhouse, M. Dolovich, G. Obminski and R. K. Wolff, *Arch. Environ. Health*, 1978, **33**, 24.
[103] B. Rudell, T. Sandström, N. Stjerneberg and B. Kolmodin-Hedman, *J. Aerosol Sci.*, 1990, **21**, S411.
[104] W. D. Bennett, K. L. Zehman, C. W. Kang and M. S. Schechter, *Ann. Occup. Hyg.*, **41** (suppl. 1), 497.
[105] P. A. Valberg and A. Y. Watson, in *Proceedings of the Second Colloquium on Air Pollution and Human Health*, ed. J. Lee and R. Phalen, Utah, 1996, p. 4–573.

to particles. For an individual, personal exposure will comprise exposure from a variety of sources, which may be of outdoor or indoor origin. Where there is a significant indoor source of particles this may have a large influence on total personal exposure levels, and remedial action may be warranted. However, it may be appropriate to focus such action upon the susceptible groups in the population. For example, ventilation could be improved, or air filters fitted, in housing for the elderly and asthmatics. Alternatively, if certain types of particle are identified as being more harmful, controls could be put in place to limit exposure to these particles (*e.g.* fit extractor hoods to gas cookers).

3 Other Issues

For the particular pollutants considered here, there is a large variation in the amount of available information on levels in the home and in the degree of confidence in measuring and monitoring techniques and in the evaluation of likely health effects. Factors such as age, social class, ethnic group, geographical area and type of dwelling may also influence the likelihood and type of health effects brought about by exposure to these pollutants. Further studies are needed to clarify some of these issues.

There are a number of other specific indoor pollutants, not reviewed above, which are also important. These include radon (a naturally occurring radioactive gas suspected of causing many cases of lung cancer[106]), organochlorine compounds (used, for example, as domestic pesticides), other biological allergens (such as cat dander), fibres (both asbestos and man-made mineral fibres[107]) and environmental tobacco smoke. Although the health effects of tobacco smoke on adults are well known, the issue of environmental tobacco smoke (ETS) in the home tends to attract little attention because it seems so obviously under the direct control of the occupants. This is not, however, the case for children, who may be unwillingly and chronically subjected to tobacco smoke in the home (or *in utero*) at a critical time of life. Understanding the impact of ETS on the health of children is thus of particular importance. Tobacco smoke contains tar droplets and a cocktail of various other toxic chemicals including carbon monoxide, nitric oxide, ammonia, hydrogen cyanide and acrolein, together with proven animal carcinogens such as *N*-nitrosamines, polycyclic aromatic hydrocarbons and benzene. ETS is known to irritate the eyes, nose and throat, and exposed babies and children are more prone to chest, ear, nose and throat infections. Also, women exposed during pregnancy tend to have lower birthweight babies. In a recent review, the Californian EPA[108] concluded that causal links have been established between the exposure of non-smokers to ETS and the following adverse conditions, and estimated their associated relative risks (RR): death from

---

[106] ECA-IAQ, *Radon in Indoor Air*, European Collaborative Action 'Indoor Air Quality and its Impact on Man', Report No 15. EUR 16123 EN, Office for Official Publications of the European Community, Luxembourg, 1995.

[107] IEH, *Fibrous Materials in the Environment*, Report SR2, Institute for Environment and Health, Leicester, 1995.

[108] EPA, *Health Effects of Exposure to Environmental Tobacco Smoke*, Final Report, California Environmental Protection Agency, Sacramento, 1997.

heart disease (RR 1.3); lung cancer (RR 1.2); nasal sinus cancers (RR 1.7–3.0); and, in children, low birthweight (RR 1.2–1.4); sudden infant death syndrome (SIDS) (RR 3.5); asthma induction (RR 1.75–2.25) and exacerbation (RR 1.6–2); middle ear infection (RR 1.62); and lower respiratory disease (RR 1.5–2). Although the relative risk for some of the conditions was small, it was noted that the diseases are common and hence the overall impact on health is potentially quite large. A current series of articles by Strachan and others on a range of disease endpoints including, for example, middle ear disease,[109] has further emphasized that ETS, especially with regard to the exposure of children, is an important issue warranting close attention.

In addition to the particular pollutants mentioned above, there are other issues related to the theme of indoor air quality and health which require a rather different approach. The first is so-called 'sick building syndrome' (SBS), which comprises a range of disparate but common conditions and is associated with certain individual workplace buildings.[110] Various causes of SBS have been proposed, including volatile organic compounds (VOCs; see above), temperature and humidity, 'dustiness/cleanliness', dust mites, mechanical ventilation and various psychological factors, but none of these alone appears to explain the syndrome adequately. Also 'multiple chemical sensitivity', the condition whereby individuals appear to show exquisite sensitivity to very low concentrations of organic chemicals, is receiving further attention, having originally spawned much interest in the USA and Scandinavia but generally being ignored elsewhere. There is also continued interest with respect to increasing trends in asthma rates in developed counties and the possible role of the indoor environment.

Policy and Research Initiatives

In the UK, the publication in 1991 of the Select Committee Report on Indoor Pollution,[111] and the Government's subsequent response,[112] served to focus attention on the importance of the indoor environment, and this momentum has been maintained in recent years. Both the Department of the Environment, Transport and the Regions and the Department of Health are commissioning research on indoor air quality directly relevant to human health effects. More widely in Europe, a number of extensive multi-centre studies are underway or awaiting final analysis (*e.g.* APHEA and EXPOLIS*). These should provide useful information for the further assessment of exposure and health impact of a number of key indoor pollutants described here.

*   APHEA: Air Pollution and Health: a European Approach; EXPOLIS: Air Pollution Exposure Distributions within Adult Populations in Europe.
[109] D. P. Strachan and D. G. Cook, *Thorax*, 1998, **53**, 50.
[110] ECA-IAQ, *Sick Building Syndrome—A Practical Guide*, European Collaborative Action 'Indoor Air Quality and its Impact on Man', Report No 4. EUR 12294 EN, Office for Official Publications of the European Community, Luxembourg, 1989.
[111] House of Commons Select Committee, *Indoor Pollution*, Sixth Report, HMSO, London, 1991.
[112] Cmnd 1633, *The Government's Response to the Sixth Report from the House of Commons Select Committee on the Environment, Indoor Pollution*, HMSO, London, 1991.

The UK National Environmental Health Action Plan (NEHAP),[113] which was published in 1996, identifies indoor air quality (IAQ) as a key area for action. This is particularly important because the UK was one of the 'pilot countries' chosen at the 1994 Helsinki inter-governmental conference on environment and health to produce the first NEHAPs, and it is thus likely that the considerations of indoor air quality contained in the UK document will carry through to other nations' NEHAPs. The UK plan provides a framework for actions by central and local government, industry and voluntary organizations to improve the environment for the benefit of health. The intention is for individuals to make informed decisions about their own homes, using appropriate, targeted information. Information is to be made available to people about the most important sources of pollutants in the home, including activities that lead to exposure, and actions which they can be taken to minimize exposure and any associated health effects. Such actions may include not smoking indoors, ensuring adequate ventilation when using a gas cooker, using water-based rather than solvent-based paints and choosing low-formaldehyde particle board or carpets. Better instructions for use of household products may also be warranted. In addition, manufacturers and suppliers of materials and furnishings are encouraged to reduce the levels of emissions from their products generally and to provide relevant product information so that people can choose to buy or specify the materials they require. The plan is to be revised and is likely to contain new initiatives relevant to indoor air quality. Targeted research in the area of indoor air quality and health is to continue in an effort to understand better the levels, sources and health effects of indoor pollutants, and to provide further knowledge about mechanisms of action and the role of mitigation procedures.

On the broader international scene, IAQ is continuing to receive attention through bodies such as the International Society for Indoor Air Quality and the major triennial 'Indoor Air' conferences. Also there is a NATO initiative on indoor air under its 'challenges of modern society' programme, and the World Health Organization is increasingly concerned about indoor air pollution in developing countries where exposures can be extremely high. Recognizing the real importance to public health of indoor air quality, the US Environmental Protection Agency recently launched a major policy and research initiative on the subject. The European Concerted Action on 'Indoor Air Quality and its Impact on Man' continues to add to its list of published reports on this theme, but with the exception of the Nordic countries, Germany and the Netherlands, policy on indoor air issues in Europe is generally poorly developed. To improve this situation, WHO Europe recently produced a document on indoor air pollution exposure assessment,[114] and is presently engaged in formulating a strategic approach to indoor air policy making. This latter publication is intended to inform and advise governments, public health authorities, and other policy makers and representatives of other sectors relevant to IAQ management on how

---

[113] DoE, *The United Kingdom National Environmental Health Action Plan*, CM3323, HMSO, London, 1996.

[114] M. Jantunen, J. J. K. Jaakkola and M. Krzyzanowski, *Assessment of Exposure to Indoor Air Pollutants*, European Series No. 78, World Health Organization Regional Office for Europe, Copenhagen, 1998.

to develop and strengthen IAQ policy in order to achieve health protection and promotion in the indoor environment. It recommends that a key strategy for the management of IAQ is the development of a comprehensive, scientifically sound and thoroughly considered 'action plan' (possibly part of a NEHAP) which should be targeted to new construction as well as existing buildings (and other indoor spaces), and should entail actions at both national and local levels. The quality of indoor air is determined by a large number of different factors and, consequently, different professions are involved in dealing with and solving indoor air problems. While central government may take the lead, industry and commerce also need to make appropriate contributions to the achievement of better indoor air quality. The role of the private sector in ensuring acceptable indoor air quality is therefore encouraged. The IAQ strategy needs also to include the assessment (and, where appropriate, the promotion or/and verification) of the 'safety' of building materials and equipment, furniture, consumer products and other materials used inside enclosed spaces. The European Concerted Action on Indoor Air and its Impact on Man is in the process of evaluating methods for the positive labelling of products which are low emitters of volatile organic compounds and has recently published a report on this topic.[115] Also there are likely in the near future to be European standards regarding emissions of formaldehyde from building materials.

One of the 'policy' issues frequently raised concerns specific standards or guideline levels for indoor air pollutants, but experiences in Germany led Seifert[116] to caution that such guideline values, if not introduced with sufficient care, may accelerate the already existing trend to solve air quality problems by litigation.

4   Conclusions

While the main focus of public concern may, for a while at least, remain on outdoor air quality, notably traffic pollution, it is clear that the indoor environment merits extra attention. This, after all, is where people spend the vast majority of their time, and the quality of the air in the home could have significant impacts in public health terms. Certainly the indoor environment has been shown to contain sources of various noxious substances. There is particular concern for potentially vulnerable or susceptible groups such as the very young, the sick (especially, perhaps, those with pre-existing cardiopulmonary disease) and the elderly, who spend a disproportionately large amount of time indoors at home. Within the scientific community there is the requirement to consider fully the role of indoor pollution in the context of total personal exposure in order to assess properly the impact of air pollution on the health and well-being of individuals and to facilitate the identification of effective control and remediation measures.

[115] ECA-IAQ, *Evaluation of VOC Emissions from Building Products—Solid Flooring Materials*, European Collaborative Action 'Indoor Air Quality and its Impact on Man', Report No 18. EUR 17334 EN, Office for Official Publications of the European Community, Luxembourg, 1997.
[116] B. Seifert, in *IEH Assessment on Indoor Air Quality in the Home, Assessment A1*, Institute for Environment and Health, Leicester, 1996, p. 324.

5 Acknowledgements

Production of the IEH report referred to in this article was funded through a research contract with the UK Department of the Environment, Transport and the Regions (DETR). Special thanks are due to Charles Humfrey, Linda Shuker, Emma Green, Christine Tuckett, Philip Holmes, Mark Taylor and Simon Short at IEH, to Linda Smith (DETR), and to all participants in the various IEH indoor air quality expert workshops. The opinions expressed here are those of the author.

Subject Index